# 22nd Topical Conference on Gas Utilization 2022



Topical Conference at the 2022 AIChE Spring Meeting and
18th Global Congress on Process Safety

San Antonio, Texas, USA
10 – 14 April 2022

ISBN: 978-1-7138-6818-7

# 22nd Topical Conference on Gas Utilization 2022

Topical Conference at the 2022 AIChE Spring Meeting and
18th Global Congress on Process Safety

San Antonio, Texas, USA
10 – 14 April 2022

AIChE
The Global Home of Chemical Engineers

**Printed from e-media with permission by:**

Curran Associates, Inc.
57 Morehouse Lane
Red Hook, NY 12571

CURRAN ASSOCIATES INC. proceedings.com

**Some format issues inherent in the e-media version may also appear in this print version.**



AIChE
120 Wall Street, FL 23
New York, NY 10005-4020

Phone: (800) 242-4363
Fax: (203) 775-5177

www.aiche.org

**Additional copies of this publication are available from:**

Curran Associates, Inc.
57 Morehouse Lane
Red Hook, NY 12571 USA
Phone: 845-758-0400
Fax: 845-758-2633
Email: curran@proceedings.com
Web: www.proceedings.com

# TABLE OF CONTENTS

# 639335 Unlocking the Decarbonization Potential of Gas-to-Chemicals for the Industry

**Abstract:**

Methane will play a key role in all the scenarios as mainly seen as a source of primary energy for transition towards net zero future. Utilization of methane for chemical production, often considered as the "Holy grail" of the 21st century but could not economically compete with the traditional petrochemistry. Sustainability value which could be brought by methane utilization for decarbonization of assets, may result in unlocking of the future for these technologies.

The industry is facing a serious threat about methane emissions. The topic was addressed at COP26 Summit and the methane emissions strategy should be published to finalize the EU Green Deal vision by the end of 2021.[1] Decentralized methane monetization solutions will significantly contribute to the mitigation of the issue.

The energy content per carbon number in methane is more than 20-30 % higher than in products of a Gas-to-Chemicals process. The smart utilization of that excess of energy shows a significant potential for decarbonization of an existing industrial platform.

There are three principally different ways to use the energy: autothermal oxidative conversion, electrification with either production of hydrogen or co-processing with $CO_2$, or utilization of selective oxidation (soft oxidants or advanced reactor and catalyst concepts to limit the overoxidation).

Autothermal conversion of natural gas offers a substantial potential for decarbonization of chemical and refining industry via avoidance of GHG emissions and $CO_2$ circularity inside assets. Co-processing of $CO_2$ in oxidative dry/wet conversion of methane is one of the examples, which provides an efficient solution for $CO_2$ utilization. Upgrading of the effluent will result in chemicals and hydrogen production. A right design of those processes may lead to the overall $CO_2$-neutral/negative gas-to-chemicals technology. An example of a such of technology could be C123 process, which is under development with participation of TotalEnergies.[2] The concept of autothermal conversion shows a great number of options and would require a development of new porous materials with dual functionalities. So, the future development in the field should go beyond the improvement of the individual process performance and must be assessed according to utilization of hydrogen, heat integration, opportunistic valorization of by-products (carbon), degree of electrification, energy efficiency, and a global contribution to decrease of the overall carbon footprint of the environment.

Electrification of methane-to-chemicals provides the possibility of conditions unattainable in a conventional thermochemical reactor. The latter will allow technical feasibility of the transformation and will result in differentiation for those technologies in term of higher selectivity, and simpler production slate. One should mention that the electrification of a traditional petrochemical process (e.g., a steam cracker) will be in a competition with the proven thermochemical technology in operation with many potentially cheaper alternative decarbonization options, e.g., optimization of energy consumption, electrification of compressors, cleaner fuel ($H_2$, $NH_3$, hythane), and $CO_2$ capturing technologies. In contrast, electrification of a methane conversion process is the vital condition to make technology feasible and economical. So, there is a very good chance that it's an electrified technology of methane conversion what will be the first process brought to the market with a realistic potential to compete globally with other decarbonization solutions. The electrified processes to produce hydrogen from methane may become particularly attractive due to credits for renewable hydrogen production in the future. Currently, one of the main $CO_2$-free alternatives for a blue hydrogen production is an electrified pyrolysis of methane.[3] The thermodynamic energy minimum to extract hydrogen from methane is about seven times lowers in comparison to the state-of-the art water electrolysis. The standard reaction enthalpy of methane pyrolysis per mole of hydrogen is also lower than the standard reaction enthalpy of the combined SMR and water-gas-shift (WGS) reaction. In theory, the SMR process also could be electrified, however the reaction produces $CO_2$ and requires $CO_2$ capturing. The main disadvantages of the methane pyrolysis are: (i) two times higher consumption of natural gas per ton of hydrogen in comparison to a steam reforming process, and (ii) a significant carbon coproduction (necessity to handle solids). In this context, valorization of carbon produced by the process is critical for the process viability and economics. Even if a premium quality of carbon would be produced by pyrolysis of methane, only a small part of this volume would be sufficient to saturate the market. However, it is important to mention that a substitution of the hydrogen supply at a typical refinery from SMR (about 40% of the overall demand in $H_2$ of a refinery) by methane pyrolysis will generate several times lower carbon amount than the current volume of the petcoke production on the same site. That means that in frame of a refinery, the quantity of the solid carbon to manage is comparable to the existing operation volumes and the option to displace the $H_2$ from SMR by methane pyrolysis is potentially feasible. One of the possible solutions to the sequestration of carbon could be its utilization as a feedstock in chemical reactions with $CO_2$, $H_2O$, $H_2S$, $O_2$ or in industrial processes. As a different look at valorization of carbon in hydrogen production from methane, one can think about leaving a part of hydrogen together with carbon in the same molecule. That could be achieved in an electrified methane non-oxidative conversion process, which results in co-production of valuable hydrocarbons together with hydrogen. Such hydrocarbons one can consider as a "carbon-sink" for hydrogen production but the "carbon-sink with a very high market size and high value at petrochemicals market, which will result in a significant additional economic value for the technology. These processes are currently under development.

One of the recent topics of a significant attention is utilization of soft oxidants. Typically, the selectivity of a typical OCM process is significantly challenged by overoxidation to $CO_2$. Other gaseous reagents such as $N_2O$, Halogens, and $S_2$ have been investigated to a far lesser extent as alternative, milder oxidants to replace $O_2$. Utilizing $N_2O$ as an oxidant could conceivably lead to enhanced $C_2$ selectivity in OCM due to its relatively mild oxidizing tendency compared to $O_2$. However, the challenge is in synthesis/regeneration of $N_2O$. In this context, utilization of sulfur is more attractive because a conventional Claus unit could recover the disulfur ($S_2$) from $H_2S$. The oxidative coupling of methane to ethylene using gaseous disulfur as an oxidant proceeds with promising selectivity.[4] However, the process co-produces a significant amount of $CS_2$, which needs to be valorized. Regarding halogens, Br-conversion technologies are the most advanced, e.g., a Br-mediated technology to transform methane to aromatics was developed by GTC.[5] The technology shows a promising carbon efficiency and carbon footprint.

One of the future focal points could become a fast-growing production of biomethane from biogas. The use of bio-methane allows significantly reducing the carbon footprint of the energy mix. However, the biomethane could be also one of the cost-competitive sources of bio-carbon for biofuel production, including bio-jet. In addition, the biomethane is produced mainly from non-edible and non-nutritional feedstocks and remains a concentrated source of clean energy, which shows many advantages relative to wet biomass in terms of investment, purification, and logistic costs. So, despite low production volumes, the development in the area remains relevant for the decentralization of economy and biofuel production.

The upcoming trends in electrification of downstream processes together with the emerging expansion in petrochemicals (crude-to-chemicals) lighter the product slate and result in necessity to valorize many methane-containing streams (off-gases). The ambitions to reduce the GHG emissions via electrification of the chemical industry could be only achieved if the on-purpose valorization of methane rich streams will take place and the off-gases will not be sent to flare or to fire heaters for energy production. Those emission mitigations may be achieved via utilization of the fuel gas in a different way, for instance, to produce useful products and clean energy vectors. The latest will result in a significant avoidance of $CO_2$ emissions from flaring of those gases, as well as additional room for $CO_2$ sink from other assets. Methane-to-chemicals may contribute in decarbonization of refinery & petrochemicals complexes via $CO_2$ avoiding from firing of off-gases, providing solutions for $CO_2$ utilization, and supplying low carbon $H_2$.

There is a very good prospects to successfully address "The Methane Challenge" with the current progress in the development of cold-wall electrified reactor, progress in the development of bi-functional stable materials, and the economic value from decarbonization of assets, which methane conversion processes could benefit. These solutions potentially allow unlocking new opportunities in sustainable transformation of light alkanes.

[1] Global voice of gas, Eliminating of methane emissions, IGU, https://www.igu.org/news/global-voice-of-gas-4/

[2] https://www.sintef.no/projectweb/c123/

[3] J. Hydrog. Energy 45 (2020), 32, 15721–15743. https://doi.org/10.1016/j.ijhydene.2020.04.100.

[4] Arinaga, A.M.; Ziegelski, M.C.; Marks, T.J. "Alternative Oxidants for the Oxidative Coupling of Methane," Angew. Chem. Int. Ed. 2021, 60, 2-16. DOI: 10.1002/anie.202012862.

[5] Unconventional Aromatics Processes PEP Report – 300, 2018, IHS

# New Developments In Methane Pyrolysis For Carbon Nanotube And H2 Production

Nanocomp Technologies Inc., Huntsman Advanced Materials

David Gailus; Steve LeViness, PhD; David Steckler, PhD; Dr. Alain Bourhis, ScD; Deepak Varshney, PhD.

# Outline

1. Executive Summary
2. Technology Overview—Floating Catalyst Chemical Vapor Deposition
3. Applications
    a. High value carbon
    b. Hydrogen
    c. Natural gas decarbonization
4. Impact of the Technology
5. Way forward.

1. Executive Summary

Floating catalyst chemical vapor deposition (FC-CVD) can be used to convert methane and other hydrocarbons present in natural gas into high value carbon and hydrogen. The value of the carbon co-product in the form of high-aspect ratio nanotubes (CNTs) can offset the process expenses, producing hydrogen that's cost competitive with SMR with a lower carbon footprint. To date this process has been used primarily for space and defense applications. Nanocomp Technologies, part of Huntsman Advanced Materials, has been working with ARPA-E to model the process at pipeline scale and other organizations are collaborating on a range of scales that can de-risk the technology. Several initial small plant applications exist in flare gas, stranded gas, and boiler gas decarbonization markets.

Nanocomp has been operating the process for over 15 years, but there are special considerations at scale that require further development or validation. For example increasing space velocity may be necessary to keep capital costs down to create attractive investment returns. This may push the operating ranges beyond the current practices and may require new materials of construction, new feed and harvest systems, and new methods to separate the hydrogen and solid carbon product. Perhaps most important will be developing products that will use the carbon material at the kiloton or megaton scale.

**Technology Overview**

CNTs have exceptional strength and conductivity but typically the tubes are relatively short, in the 10s of micron range, with diameter to length aspect ratios in range of 1:100 to 1: >5000. While nanotube are known to be "sticky" due to their enormous surface area, CNT's of these aspect ratios do not have enough lateral contact to start to mechanically transfer load to each other in shear or tension. As a result they act like powders. When the aspect ratios increase to 10,000 and above, the CNTs bundle and entangle forming networks that can transfer force. CNTs can then be spun into yarns or directly form sheets without a binder. CNT's with these high aspect ratios are enabled by the floating catalyst CVD method.

Floating catalyst CVD to make spinnable carbon was first reported by Gary Tibbetts at GM Research Center in 1984. This was seven years before Sumio Iijima describe what we now call nanotubes. Tibbetts' patents have long since expired but the work was carried on and potentially rediscovered by Ya Li-Li and Alan Windle at Cambridge University in the early 2000s, who published and patented a large body of work on the topic. Since then many labs around the world have replicated the process on small scale. Nanocomp developed the process to product sheet and yarn materials, primarily for DoD and space applications such as damage tolerance pressure vessels for hydrogen transport, EMI and micrometeoroid protection for spacecraft and conductors for deployable antennae. The process can use a wide variety of carbon sources. The trade name adopted for these materials is Miralon™ with several designations depending on form, composition and post processing steps.

The process starts with injecting a nano-metal catalyst, in this case, nano-iron. This nano-iron can be introduced as an aerosol of preformed nanoparticles. More commonly the nano-iron is formed in situ by the thermal decomposition of an organometallic, in this case ferrocene. Ferrocene decomposes between 350°C and 750°C depending on the atmosphere. Ferrocene decomposes into iron vapor and cyclopentadiene. The iron vapor is well below the boiling point of iron and it immediately starts to condense into droplets. When the droplets reach 1 to 10 nanometers in diameter they are injected into a reactor along with a hydrocarbon source and a partial pressure of hydrogen and quickly heated to over 1100°C. The iron particles oversaturate with carbon and start to grow carbon nanotubes. The tubes grow very quickly, reaching lengths up to 1mm in a few seconds. These high aspect ratio CNTs bundle and entangle to form an elastic network that can be harvested as a low density solid. The initial partial pressure of hydrogen suppresses soot formation, favoring the formation of more structured carbon.

If methane is used as the feedstock, the byproduct is hydrogen. Single pass conversion efficiency can be greater than 80%.

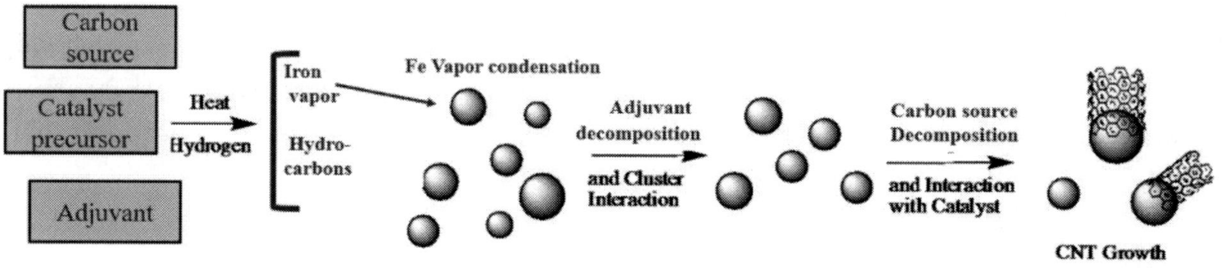


Figure 1. Process sequence for floating catalyst CVD formation of carbon nanotubes.

Carbon nanotubes are produced in bulk for lithium ion batteries, supercapacitors and conductive filler applications, typically using fluidized bed, fixed bed or continuous belt substrate methods. As mentioned CNTs are typically short, making them more valuable for electrical applications, less so for mechanical applications. In contrast the high aspect ratios made possible with the floating catalyst method allow enough CNT-to-CNT interaction to transfer load, opening up a wide variety of structural applications. Figure two shows yarns and sheet material consisting of CNT yarn and sheet material produced without a binder. This is possible because of the bundled and entangled network that is naturally formed as the CNTs grow, drift and coalesce from the gas phase.

## High Value Carbon—Nature and Properties

Miralon is currently produced in sheet, yarn and pulp formats. The properties are controlled by the reactor process conditions, the harvesting method and the post processing steps. Interestingly, the hydrocarbon source also influences the product properties in spite of the fact that the high temperatures quickly break down organic inputs into a similar spectrum of compounds. Highest strength materials tend to be made from alcohols and aromatics. Methane and other alkanes are most productive and have the highest hydrocarbon to carbon conversion efficiencies.

Miralon sheet materials as produced have tensile properties between 60MPa and 300MPa highly dependent on the incoming feedstock and the degree of densification. In sheet format, the high aspect ratios CNTs lie primarily in-plane which gives the material highly anisotropic mechanical and thermal properties. In plane thermal conductivity is typically 20 to 100 W/mK while the thru-thickness conductivity can be as low as 0.03 W/mK. Z-axis mechanical properties are also much lower, potentially by two orders of magnitude.

Mechanical applications will typically require an organic matrix to bind the CNTs together giving the composite system with both tensile and compressive strength. Binders can be thermoplastics, thermosets or inorganics. The challenge it to create interaction with the CNTs at the nano scale to enhance load transfer. High volume applications for this material are targeting the tensile strength of structural steel, roughly 400MPa.

Miralon yarns are produced with emphasis on alignment of the CNTs along the axis of the fiber to give maximum uniaxial mechanical and electrical properties. The highest strength yarns produced to support NASA programs approach 4 GPa, in the range of the strongest carbon fiber. Alignment requires control within all steps of production from the CNT synthesis and agglomeration to the harvesting and post process stretching and densification. Nanocomp has developed post processing steps that can increase the tensile properties of the yarn by a factor of four. It is important to note that the highest strength materials are not made purely from methane and the conversion and volumetric efficiencies of the reactors producing yarn are poor compared to those producing sheet and pulp materials.

Miralon pulp is made by untangling the CNT bundles in a series of post processing steps. CNTs have a very high affinity for each other so care is taken to design the process to specifically tease the CNT bundles apart, separating them without damage. If the aspect ratios of the individual CNTs is retained, the pulp can be recast or compounded to retain some of the extraordinary properties. The current method starts with a manufacturing intermediate in the form of rolled, low density entangled Miralon material harvested directly from the reactors. This then goes through a series of size reduction steps all based on hydrodynamic shear rather then cutting or milling that could damage the CNTs. The current process is water based and the material can be dewatered and dried or transferred directly into an organic solvent or matrix by solvent exchange. The pulp has some extraordinary properties. Added to polymers in concentrations of less than 0.01% can make them electrically conductive without changing other properties. In higher concentrations they can complement other structural materials improving their mechanical properties and fire resistance. They can also be the basis of structural materials with the addition of matrix polymers.

**Hydrogen Production**

When natural gas is used at the carbon source, the byproduct is hydrogen. As mentioned earlier a small amount of hydrogen is needed to promote the formation of structured carbon, but this hydrogen is not consumed and can be recovered and recycled. Natural gas to carbon and hydrogen conversion efficiency can be greater than 90% single pass, with 80% as a practical target. The hydrogen enriched residual gas can be separated or recycled or fed into cascading reactors to improve the overall system efficiency. The value of a cascading architectures is the reuse of some of the thermal energy transferred with the hydrogen.

The target of the initial ARPA-E funded effort was to evaluate FC-CVD comparing the economics and environmental impact to steam methane reforming (SMR). SMR is a highly productive process with much of the hydrogen coming from water in addition to the source methane. The downside of SMR is the CO2 byproduct. FC-CVD does not create CO2, instead it removed the carbon as a solid. However it does require thermal energy both to heat the incoming feedstocks and to drive the endothermic reaction. The carbon footprint of the FC-CVD process depends on the source thermal energy. The thermal energy could come from electricity from the grid or from combustion of natural gas or from combustion of roughly half of the hydrogen product. If the grid electricity is from renewables, the process has a very low carbon footprint compared to SMR. The decision to use hydrogen for thermal energy is an economic one and depends on the value of the hydrogen at the location where it is being produced. The if the hydrogen has high value, burning natural gas with carbon capture could be the most economical and environmentally benign choice. Recovering at least part of the thermal energy is also beneficial but will depend on the design of the reactor and harvesting system.

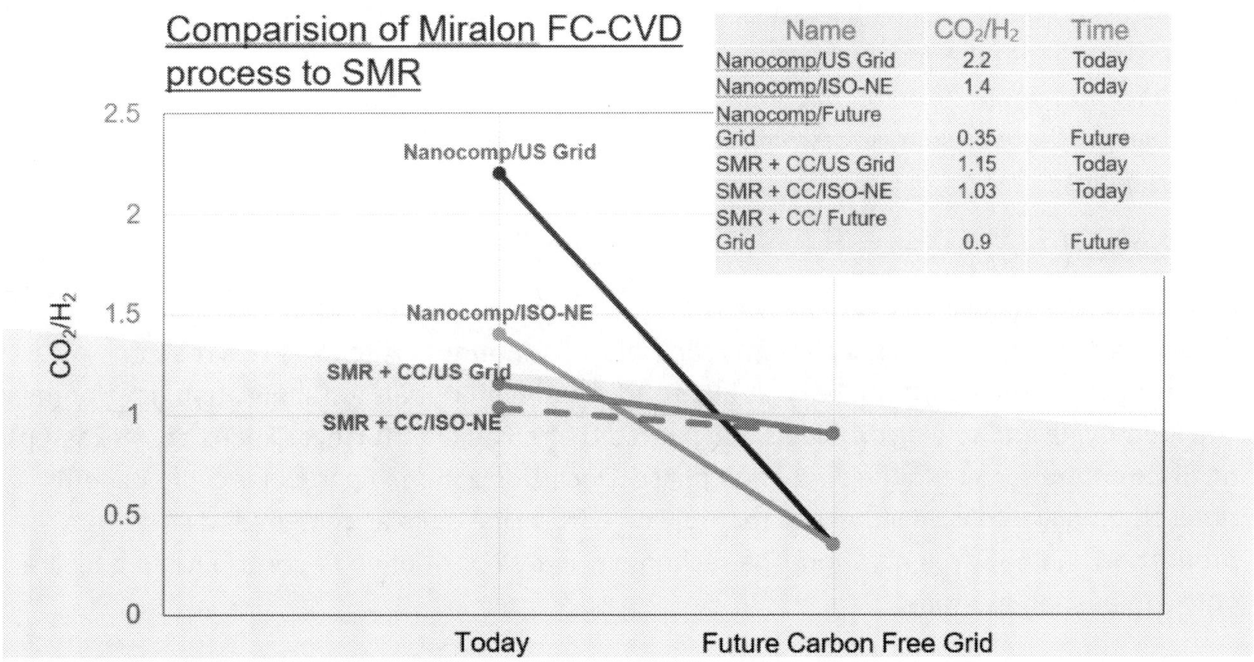


Figure 3. Comparing FC-CVD with SMR

**Impact of the Technology**

The initial interest for FC-CVD focused on the carbon. High strength, light weight materials are of great interest to the defense and aerospace community. From the interaction with and investments from ARPA-E, several new opportunities have opened up. ARPA-E's original interest was to evaluate the process at the megaton scale, looking for ways to make competitive hydrogen by offsetting the process costs with a high value co-product. Work started in 2018 with a combination of modeling and experimentation to improve conversion efficiency and space velocity and create conceptual designs for a megaton scale reactor to allow for economic analysis. Using the H2Fast NREL modeling tools to estimate the hydrogen cost, Nanocomp was able to demonstrate competitive hydrogen at a price for the carbon of <$0.40/kg. One of the challenges is finding large markets for the megaton quantities that would be produced. This work is ongoing.

Spinning out from this work, COSIA, the Canadian Oil Sands Innovation Alliance collaborated on a slightly different objective. They were interested to in seeing if FC-CVD could be used to decarbonize natural gas that they use for thermal energy needed to generate large amounts for steam for oil sands production. Since the FC-CVD is a thermal process, the COSIA derivation of the process offers some advantages. Waste heat from the process could be recovered to make steam. The energy to concentrate and purify hydrogen from residual gas was not essential to the objective, potentially reducing the capital requirements. In this analysis, once again the

GHG footprint depends on the source of thermal energy, whether combustion on site or use of electricity from the grid and the source of that grid's energy.

Other applications being investigated include flare gas remediation. This has some particular challenges, primarily based on logistics. In evaluating this over the past year, we have demonstrated that the process is relatively robust to non-methane components in flare gas including nitrogen, $H_2S$, water vapor and high molecular weight hydrocarbons. However, portability of large-scale reactors will require innovations on the design and materials of construction.

**Way Forward**

Huntsman is developing a large-scale system to shift production of carbon from the 10s of kilograms per year scale to the kiloton/year scale. The intention is to have this reactor operating in a demonstration mode in 2023 and creating materials for application development and identifying and solving problems anticipated that could limit further expansion. Next steps on decarbonization are in the discussion phase. Decarbonization in particular could benefit other thermally intensive industries that rely on natural gas such as cement and steel production and power generation. The technology is well positioned to contribute to a future post combustion economy.

SPRING22
+18TH GCPS
A Joint AIChE and CCPS Meeting

# Extraction of Chloride Salts from LPG Product Streams Using a Proven Extraction Technology

**Evan Lowry, Grant Jordan, Jonathan Gonzales, Sharon Maydak, Yanxiao Yuan**
**Pentair Oil & Gas Separations**
**4301 W. Davis St., Conroe TX**
**evan.lowry@pentair.com**



Prepared for Presentation at
American Institute of Chemical Engineers
2022 Spring Meeting and 18th Global Congress on Process Safety
San Antonio, TX
April 10 – April 14, 2022

# Extraction of Chloride Salts from LPG Product Streams Using a Proven Extraction Technology

**Evan Lowry, Grant Jordan, Jonathan Gonzales, Sharon Maydak, Yanxiao Yuan**
**Pentair Oil & Gas Separations**
**4301 W. Davis St., Conroe TX**
**evan.lowry@pentair.com**

**Keywords:** LPG, olefins, catalyst poisoning, water-wash

## Abstract

Salt cavern storage is a common method used for holding processed liquefied petroleum gases (LPGs) such as propane and butane prior to sale. By high pressure pumping of high TDS brine into the salt cavern via an injection well, the hydrocarbon product can be displaced and recovered from the salt dome and is subsequently transported to point-of-sale. One challenge with the recovery of stored hydrocarbon product is the removal of entrained and dissolved brine and salts from the LPG which are regarded as contaminants. Often, simple knock-out drums are installed at the wellhead but these devices suffer from carry-over and may fail to meet the required specifications. A large southern petrochemical and olefins plant was experiencing issues with meeting the required specification on their LPG recovered from salt domes. The requirement of 1 mg/L of inorganic chlorides and no free water was not achievable with a simple knock-out drum. Using in-house sophisticated modeling techniques, Pentair was able to provide a design proposal for a transient salt and brine recovery unit for the recovered LPG stream based on the Pentair Polarex® technology. The unit was installed and brought online in mid-December 2020. Samples taken upstream and downstream of the extraction system show that the required extraction and separation was achieved by utilizing a counter-current recycle configuration. The plant was able to minimize water usage while meeting the desired specifications for the recovered n-Butane product. The developed modeling techniques showed good agreement with the operating units with minor differences attributable to transient operating conditions.

## 1 Introduction

Feedstock contamination in catalytic dehydrogenation and cracking units is a persistent issue in the petrochemical industry. Many catalysts are highly sensitive to the amount of metal ion contaminants that enter with the feed. Due to the high temperatures that are employed in fluidized and fixed catalytic cracking units for olefin production, metallic ions have the tendency to replace or occupy active site in the catalyst particles [1]. With increased production of natural gas, a large amount of LPG fractions is utilized for production of propylene and ethylene, common petrochemical and plastic precursors. Most catalysts for dehydrogenating light alkane feedstock into olefins are composed of mesoporous zeolite with optimized amounts of aluminum and silica such as ZSM-5 [2]. Among common contaminants, sodium and chloride are identified in the literature as producing deleterious effects on catalysts. Sodium ions entering with the feed can cause catalyst sintering and loss of surface activity via de-alumination of the matrix [3]. Tangstad et al. (1999) found that sodium chloride results in catalyst deactivation for zeolitic FCC catalyst [4]. Senter et al. (2021) found that chloride ions can reactivate other metal contaminants, like nickel, which can change the catalyst properties or reduce the catalyst activity [5]. Both sodium and chloride are commonly found in LPG and butane that are stored in underground salt caverns. The hydrocarbon product is contacted with high saline brine during the storage process. Some of the brine and salts remain entrained with the hydrocarbon when it is ready to be processed. As a result, these contaminants can find their way to the catalytic reactor of the olefins unit prior to petrochemical processing. Shutting down the refinery for a turn-around to change out the fouled catalyst can cost millions of dollars in lost productivity.

# 2 Background

## *2.1 Process Conditions*

A southern petrochemical and olefins plant acquired a large amount of butane stored in underground salt caverns. The butane was planned for conversion to olefins, which were then shipped to a petrochemical facility as feedstock. The olefins catalytic reactor was very sensitive to chloride contamination and therefore a 1 ppmw limit was set for the incoming butane. The produced butane was expected to have some chlorides due to brine entrainment from the salt cavern. As a result, the pipeline operators began evaluating solutions to remove the chloride prior to pipeline transport into the olefins reactor.

The butane product was stored in a salt dome that was cased to about 2,000 ft below surface level. Primarily the product was composed of normal butane with some butadienes and isobutane as shown in Table 1. The wellhead was designed to operate at 600 psig and 80 °F with a projected flowrate of 7,000 BPD (205 USGPM).

**Table 1. Composition of butane mixture in storage cavern**

| Species | Composition | |
|---|---|---|
| Propane | vol% | 1.20 |
| Isobutane | vol% | 35.91 |
| n-Butane | vol% | 40.33 |
| 1,2-Butadiene | vol% | 0.21 |
| 1,3-Butadiene | vol% | 0.19 |
| 1-Butene | vol% | 6.92 |
| Isobutene | vol% | 5.15 |
| cis-2-Butene | vol% | 2.42 |
| trans-2-Butene | vol% | 3.56 |
| Pentane+ | vol% | 4.11 |
| Sulfur | ppmv | 200 |

## *2.2 Pentair Polarex® System*

The cavern brine was known to contain between 20-35 wt% sodium chloride that may be entrained in the butane product as it is displaced from the cavern. Because of this, a particle filter and liquid-liquid coalescer were placed in front of the water wash system at the wellhead. The particle filter reduced the amount of solids entering the system while the inlet brine coalescer removed the free (non-soluble) brine from the butane product. Sophisticated process modeling indicated that between 260-450 ppmw of brine would be soluble in the butane mixture at process conditions. Based on this estimate, a two-stage Pentair Polarex® water wash treatment system was designed based on in-house process simulations to treat the butane and prevent carry-over of water and brine to the pipeline. A total freshwater injection rate of 3 GPM was used for design and commissioning of the system.

After finalizing the design, the equipment was fabricated, installed, and connected to the cavern. A general layout is presented in Figure 1. Images of the installed system are shown in Figure 2 and Figure 3.

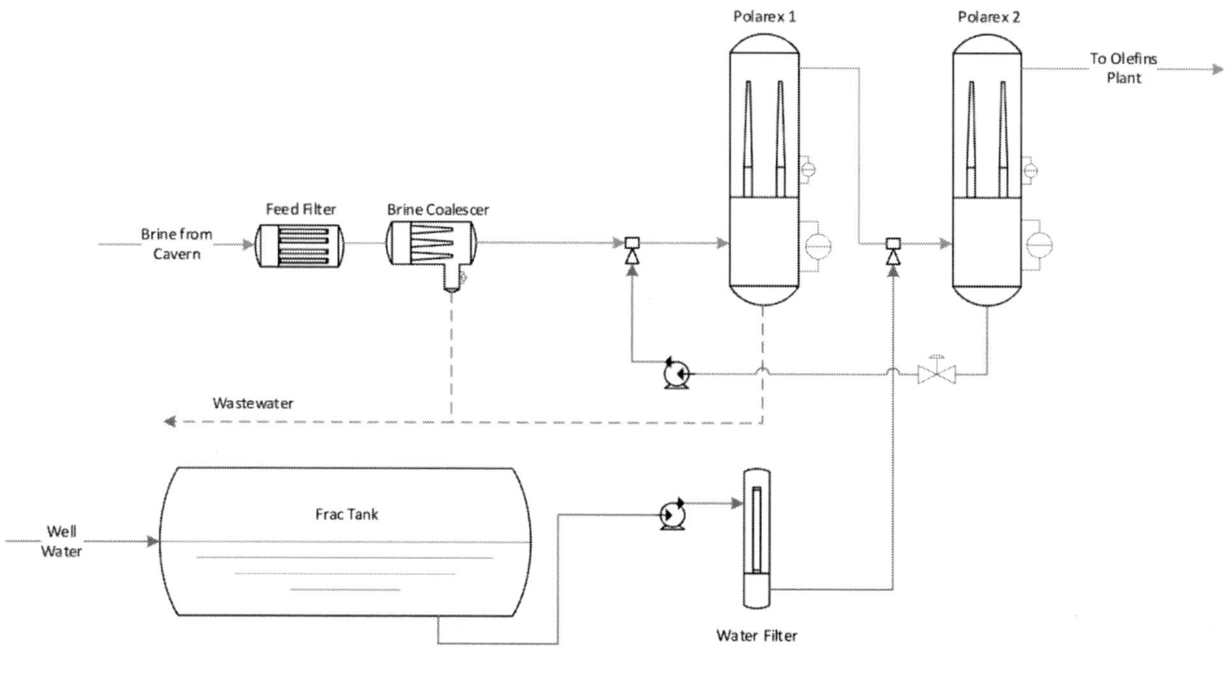


**Figure 1. General process layout**

**Figure 2. Polarex® system on-site prior to piping connections**

**Figure 3. Injection water filtration system**

# 3 Results

## *3.1 Butane Sampling*

After start-up and commissioning, samples of the butane mixture were taken from the inlet and outlet of the Polarex® system to ascertain system performance. Cylinders with butane were collected in the field under pressure and returned to the lab for analysis by the conductivity method. All glassware was initially triple washed with de-ionized water, washed with methanol, and dried for 24 hours in an oven at 400 °F. Each cylinder was depressurized by sparging the contents of the cylinder through a glass frit and into a 500 mL glass column filled with 100g of de-ionized water. The water was weighed before and after contacting the contents of the cylinder for measurement error determination. The wash water was recovered from the glass column and the electrical conductivity was measured using a high precision conductivity probe having a resolution of 0.1 μS/cm. The conductivity value was recorded and used to calculate the total salt content in the butane sample after subtracting the background conductivity (if any) of the original de-ionized water. The conductivity method does not distinguish between salt ion species, as a result, all salt concentrations were reported as sodium chloride (NaCl).

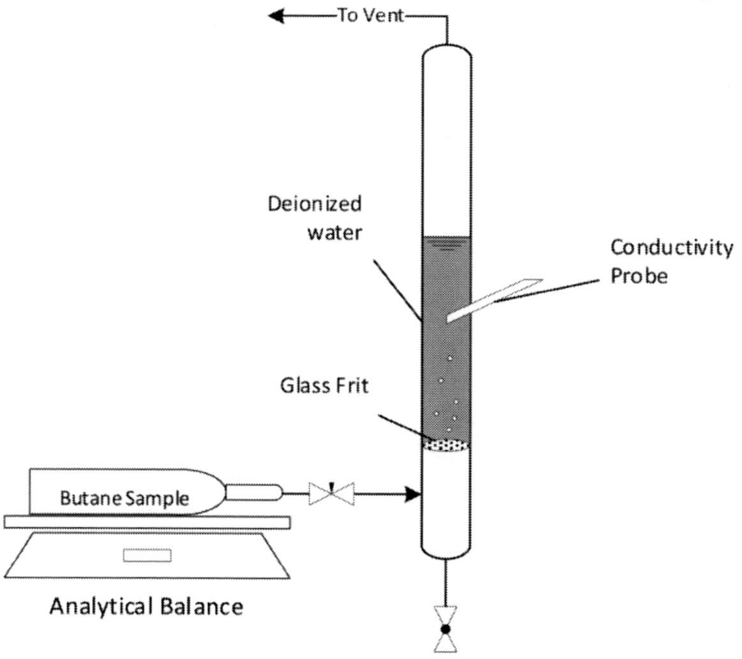


**Figure 4. Analytical set-up for laboratory chloride analysis**

**Table 2. Results of butane sampling analysis after system startup**

| Date | Time | Sample ID | Total Salt in Butane (ppmw) | Est. Error (%) |
|---|---|---|---|---|
| 1/11/2021 | 9:45 AM | Outlet | 1.17 | 10.3% |
| 1/11/2021 | 9:55 AM | Outlet | 0.79 | |
| 1/11/2021 | 10:05 AM | Outlet | 1.05 | |
| 1/11/2021 | 9:50 AM | Inlet | 5.11 | |
| 1/11/2021 | 10:00 AM | Inlet | 1.89 | |
| 1/11/2021 | 10:10 AM | Inlet | 2.03 | |

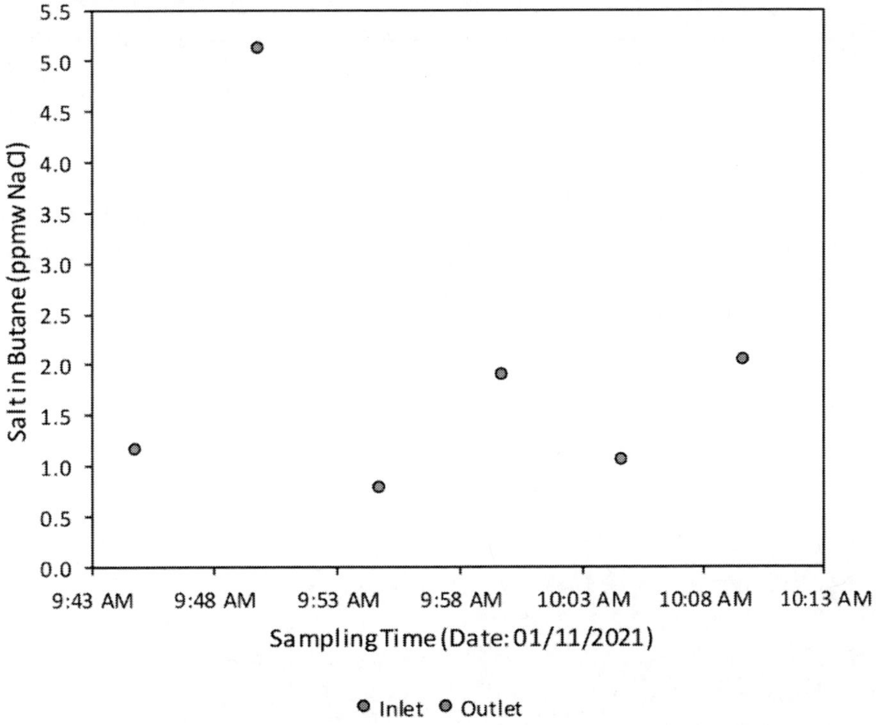


**Figure 5. Results of butane sample analysis after system startup**

### *3.2 Water Analysis*

One challenge that emerged at the plant during the start of the project was the injection water quality. A non-potable water source was available on-site from a water well. This water was originally used until rapid filter fouling was observed on the water filter. Samples were taken of the well water source on-site to look for contamination. Additionally, other water sources were evaluated including nearby municipal water, water from the olefins plant, and water that was trucked to the site via tanker trailer. Total suspended solids were analyzed in addition to the chemical profile of the well water.

**Table 3. Results of water sample analysis after system startup**

| Sample | Sample Date | Analysis Date | Volume (mL) | Solid Concentration (mg/L) | TDS (ppmw as NaCl) | pH |
|---|---|---|---|---|---|---|
| Municipal Water Source | 1/14/2021 | 1/20/2021 | 500 | 1.0 | 230.4 | 8.68 |
| Plant Water | 1/14/2021 | 1/20/2021 | 500 | 2.3 | 695.8 | 8.13 |
| Purchased Tanker Water | 1/14/2021 | 1/20/2021 | 500 | 0.8 | 245.8 | 8.21 |
| Well Water | 1/11/2021 | 1/12/2021 | 200 | 47.0 | 849.9 | 7.16 |

The water analysis revealed that the well water was contaminated with high levels of solids and dissolved ionic materials (salts). Scanning electron microscopy - energy dispersive spectroscopy (SEM-EDS) analysis indicated that the contaminants were primarily iron based oxides and sulfides. The water salinity was also high.

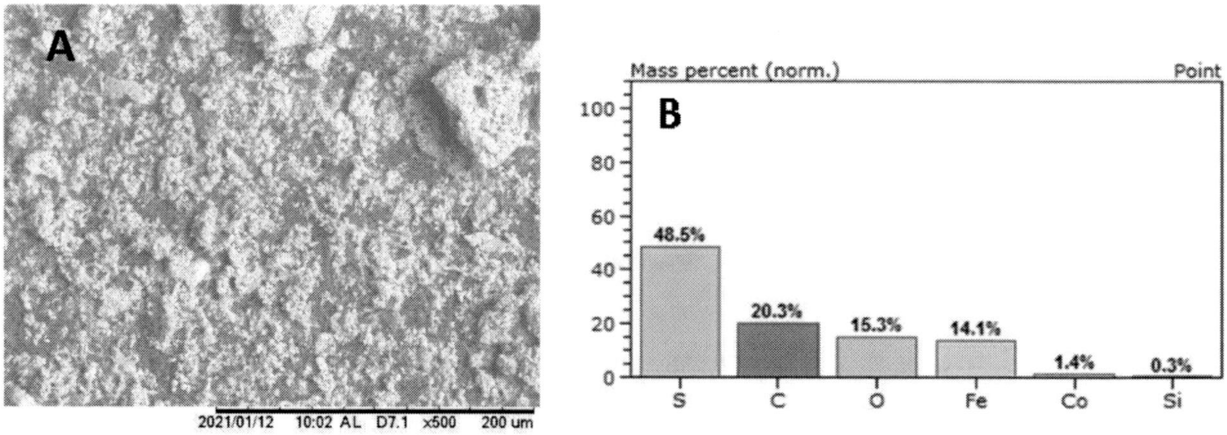


**Figure 6. SEM-EDS results of well water contaminants showing (A) SEM image at 500X magnification, and (B) EDS results of contaminant profile**

### *3.3 Process Simulations*

The water analysis and butane samples were combined with the process simulation model to help evaluate possible implications of the water quality. The butane sampling matched well with the process simulation data considering the TDS measured in the injection water from the original injection well source. Based on the model, the plant chose to utilize a municipal water source with lower TSS and TDS in order to allow for larger operational range and better outlet butane quality. By switching to a cleaner water source, the plant operation was able to maintain a consistent outlet quality and lower filter change-out costs.

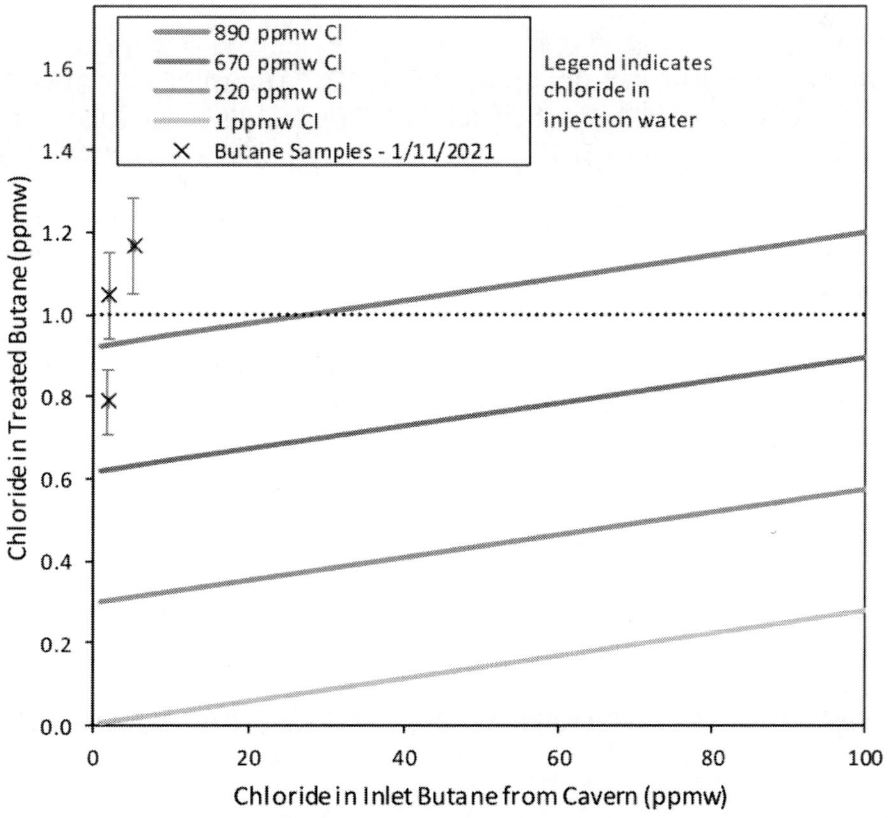


**Figure 7. Process simulation results indicating the impact of water TDS on outlet butane quality (lines). Butane sample analysis is also shown for reference.**

# 4 Conclusions

Chloride contamination in butane and other compressed hydrocarbon products is a persistent issue that must be addressed when utilizing product stored in underground salt domes. Ionic contamination can cause catalyst fouling in olefin reactors, resulting in costly change-outs and turnarounds. A two-stage water wash system was design based on the Polarex® system to provide removal of brine carry-over and soluble chlorides from stored butane mixture prior to processing at an olefins plant in the southern U.S. The system was shown to provide excellent removal of chloride salts through sampling and lab testing. Several key conclusions emerged from the case study of the operating installation:

- The two-stage Polarex® water-wash system was able to produce reactor quality butane with no free water and less than 1 ppmw of chlorides
- Sophisticated process simulations were able to aid in designing a custom, modular system with very short lead time
- Counter-current recycle configuration allows for reduced solvent usage
- Wash-water source is important when considering key specifications for the product. TDS and TSS in the wash water can cause rapid filter fouling or off-spec product in some cases

## 5 References


1. Sandoval-Díaz LE, Ruíz-Cardona YS, Trujillo CA. Amorphization of USY zeolite induced by sodium chloride and high temperature steaming. *Microporous Mesoporous Mater*. 2016;224(December):168-175. doi:10.1016/j.micromeso.2015.11.021

2. Bari Siddiqui MA, Aitani AM, Saeed MR, Al-Khattaf S. Enhancing the production of light olefins by catalytic cracking of FCC naphtha over mesoporous ZSM-5 catalyst. *Top Catal*. 2010;53(19-20):1387-1393. doi:10.1007/s11244-010-9598-1

3. Bai P, Etim UJ, Yan Z, et al. Fluid catalytic cracking technology: current status and recent discoveries on catalyst contamination. *Catal Rev - Sci Eng*. 2019;61(3):333-405. doi:10.1080/01614940.2018.1549011

4. Tangstad E, Myhrvold EM, Myrstad T. A study on the effect of sodium chloride deposition on an FCC catalyst in a cyclic deactivation unit. *Appl Catal A Gen*. 2000;193(1-2):113-122. doi:10.1016/S0926-860X(99)00414-7

5. Senter C, Mastry MC, Zhang CC, Maximuck WJ, Gladysz JA, Yilmaz B. Role of chlorides in reactivation of contaminant nickel on fluid catalytic cracking (FCC) catalysts. *Appl Catal A Gen*. 2021;611(August 2020):117978. doi:10.1016/j.apcata.2020.117978

# 639364 Direct Methane Activation By Atomically Thin Platinum Nanolayers on Two-Dimensional Metal Carbides

**Abstract:**

Efficient and direct conversion of methane to value-added products has been a long-term challenge in shale gas applications. Here, we show that atomically thin nanolayers of Pt with a single or double atomic layer thickness, supported on a two-dimensional molybdenum titanium carbide (MXene), catalyse non-oxidative coupling of methane to ethane/ethylene ($C_2$). Kinetic and theoretical studies, combined with in-situ spectroscopic and microscopic characterizations, demonstrate that Pt nanolayers anchored at the hexagonal close-packed sites of the MXene support can activate the first C–H bond of methane to form methyl radicals that favour desorption over further dehydrogenation and thus suppress coke deposition. At 750 °C and 7% methane conversion, the catalyst runs for 72 hours of continuous operation without deactivation and exhibits >98% selectivity towards $C_2$ products, with a turnover frequency of 0.2–0.6 $s^{-1}$. Our findings provide insights into the design of highly active and stable catalysts for methane activation and create a platform for developing atomically thin supported metal catalysts.

# 639393 Oxidative Coupling of Methane over Supported Mn-Na-W Catalyst: Influence of Preparation Method and SiC Crystal Phase

**Abstract:**

## 1. Introduction

The oxidative coupling of methane is regarded as a potential one-step route for $C_2$ chemicals from natural gas. However, its industrial application relies on high $C_2$ yields (>30%) with prolonged times on stream and thus, the development of catalysts that can fulfill such requirements is key. Over the years, the high selectivity of Mn-Na-W/$SiO_2$ has been systematically reported. These 3 metals create a system that converts amorphous $SiO_2$ into cristobalite which selectively allows the coupling of methyl radicals to form $C_2$ compounds. Nonetheless, the high exothermicity of the reaction typically leads to deactivation and the inclusion of SiC in the support could help dissipate the heat. Although there are some works available wherein SiC is the support of choice for the OCM, there is no study on partially replacing $SiO_2$ for SiC and the effect type of SiC crystal phase. The present work outlines a novel SiC-$SiO_2$ support synthesized by spray drying to extend the lifetime compared to the benchmark $SiO_2$-supported catalyst prepared by wetness impregnation. Through the inclusion of SiC, reaction heat is effectively dissipated while attaining high $C_2$ selectivity. The effect of SiC crystal phase is also evaluated, with $\alpha$+$\beta$ SiC fostering the formation of a convenient SiC-cristobalite core-shell structure.

## 2. Experimental

Mn-Na-W catalysts with a $SiO_2$-SiC supports were prepared by spray-drying, a novel, cost-effective and scalable process with perfect control of the properties of the final solid. Prior to spray-drying, a slurry of precursors containing the metals, $SiO_2$ and SiC was prepared and ball-milled. The slurry was then spray-dried with compressed air acting as carrier gas through a two-fold nozzle and droplets were dried in the chamber at 200 °C. Finally, catalysts were calcined at 800 °C for 6 h with a ramp of 10 °C $min^{-1}$. Catalysts with various SiC crystal phases ($\alpha$, $\beta$, $\alpha$+$\beta$) were prepared. A pure $SiO_2$-supported Mn-Na-W catalyst prepared by impregnation was considered as benchmark. Catalysts were thoroughly characterized and tested in a high-throughput 16-fold parallel reactor.

## 3. Results and discussion

The evolution of $CH_4$ conversion with time on stream for standard Mn-Na-W/$SiO_2$ catalyst prepared by impregnation (IMP $SiO_2$) and spray-dried Mn-Na-W/$SiO_2$-SiC (SD $SiO_2$-SiC) at 800 °C. IMP $SiO_2$ presents a clear descending trend, with 66% residual conversion at the end of the run (70 h). This observation antagonizes with the remarkably stable behavior observed in the case of SD $SiO_2$-SiC, showing a little drop in $CH_4$ conversion over time on stream even after 70 h. Based on $H_2$-TPR and SEM characterization, a more uniform distribution of active phases and SiC across the particles was observed for SD $SiO_2$-SiC compared to IMP $SiO_2$. From the Raman spectra of fresh and used catalysts, the presence of surface α-cristobalite could be confirmed for both fresh catalysts, which was transformed into quartz after reaction, in line with PXRD observations. Characteristic $Na_2WO_4$ (identified as a key species for OCM) peaks in both fresh and used IMP $SiO_2$ samples could also be observed, while for the SD $SiO_2$-SiC, only a meagre $Na_2WO_4$ peak could be identified in the fresh sample, due to the transition of these species into $MnWO_4$. In fact, the inclusion of SiC into the support led to the presence of $Mn_7SiO_{12}$ and $MnWO_4$ in almost absence of $Na_2WO_4$, suggesting that the improved thermal conductivity of SD $SiO_2$-SiC accelerates the dynamics of active phase transfer between $Na_2WO_4$-$Mn_7SiO_{12}$ to $MnWO_4$-$Na_2O$ which results in an enhanced stability.

When studying the effect of SiC crystal phase, the catalyst supported onto α+βSiC (SD $SiO_2$-a+bSiC) performed reasonably better than its more porous analogue (SD $SiO_2$-βSiC) in terms of $CH_4$ conversion. The bulk nature of the SiC was confirmed by TG analysis, where βSiC was converted to $SiO_2$ in greater proportions. Therefore, a bulk oxidation of SiC and more uniform distribution of SiC and $SiO_2$ was anticipated for the catalyst supported onto $SiO_2$-βSiC compared to SD $SiO_2$-α+βSiC, where a SiC-enriched core a $SiO_2$-rich shell is hypothesized based on XPS experiments and is also further confirmed from SEM analysis. As a result, the increased thermal conductivity explains the observed enhanced OCM activity of SD $SiO_2$-α+βSiC.

## 4. Conclusions

In the present work, the positive effect of incorporating SiC as a support ingredient was corroborated compared to the benchmark Mn-Na-W/$SiO_2$ catalyst. SiC inclusion endows the catalysts with more phase uniformity and increased thermal conductivity besides presumably altering the dynamics of the active sites. At the same time, the importance of SiC crystal phase in the formulation of the support was also confirmed, with an α-SiC component preventing the major oxidation of SiC to $SiO_2$, as observed in the β counterpart.

**Acknowledgements**

The authors acknowledge financial support, resources and facilities provided by the King Abdullah University of Science and Technology (KAUST).

# 638947 The Blue and Green Energy Translation

**Abstract:**

The presentation will detail the current trends in green and blue Hydrogen, ammonia and methanol for the new energy transition from fossil to blue and green energy sources. We will also discuss the production technology options for each product. The presentation will also discuss the advantages of Solid Oxide Electrical Cell technology (SOEC) over Alkaline electrolysis for the production of green H2.

# VAPOR CLOUD EXPLOSIONS AT NIL WIND

*Filippo Gavelli, Ph.D., P.E.*
*Bryant Hendrickson, P.E.*
*Blue Engineering and Consulting*

**Abstract**

Current regulations for siting of LNG facilities specify the range of wind speeds to be considered as those which result in longer predicted downwind dispersion distances than other weather conditions at the site at least 90 percent of the time. Based on these requirements, the bounding cases in most LNG facility siting studies are based on wind speeds of 1 or 2 m/s, depending on the wind data for each site. Since the consequences of vapor dispersion scenarios, including the size of flammable vapor clouds that may result in explosions, are generally assumed to become worse as the wind speed decreases, the argument has been raised that a worst-case study should consider wind speeds down to zero (i.e., 'nil-wind'). However, before regulations are potentially modified to include broader requirements, the actual effect of wind speed on flammable hazards should be carefully evaluated and include consideration of the frequency of occurrence of nil-wind conditions.

BLUE led a PHMSA-sponsored research project to evaluate the effect of nil-wind conditions on flammable hazards for a wide range of potential scenarios at LNG facilities. The evaluation was performed both under the current prescriptive regulations as well as the risk-based siting methodology outlined in NFPA 59A-2019.

**Background**

Federal safety regulations for siting, design, construction, operation and maintenance of LNG facilities are codified in 49 CFR Part 193. Siting requirements are given in Subpart B, with ambient conditions for flammable vapor dispersion specified in section 193.2059(b)(2), which reads:

Dispersion conditions are a combination of those which result in longer predicted downwind dispersion distances than other weather conditions at the site at least 90 percent of the time, based on figures maintained by National Weather Service of the U.S. Department of Commerce, or as an alternative where the model used gives longer distances at lower wind speeds, Atmospheric Stability (Pasquill Class) F, wind speed = 4.5 miles per hour (2.01 meters/sec) at reference height of 10 meters, relative humidity = 50.0 percent, and atmospheric temperature = average in the region.

Since the consequences of vapor dispersion scenarios are generally believed to become worse as the wind speed decreases, it has been argued that a worst-case study should consider wind speeds down to zero. This argument was supported by a 2017 report by the Health and Safety Executive (HSE), which reviewed several large vapor cloud explosion accidents that occurred in industrial facilities around the world and found that "a high proportion of vapor cloud incidents occurred in nil/low wind conditions" [1].

In September 2020, the U.S. Department of Transportation's Pipeline and Hazardous Materials Safety Administration (PHMSA) awarded Blue Engineering and Consulting (BLUE) a research project titled "Vapor Cloud Explosions at Nil Wind". The scope of the project was to evaluate how "nil wind" conditions may affect vapor cloud explosion (VCE) hazards in LNG facilities and to present the findings so that PHMSA staff can make an informed decision regarding possible changes to regulatory requirements. This paper gives a summary of the project findings; interested readers can find additional information on the project's page, at https://primis.phmsa.dot.gov/matrix/PrjHome.rdm?prj=922.

## Nil-Wind Conditions

The first step in addressing the effects of nil-wind conditions is to define what constitutes "*nil-wind*". The HSE report attempted to define "nil/low-wind" by recognizing that such conditions "*develop in stably stratified atmospheric conditions and are easily recognised. The density gradient near the ground is sufficient to suppress turbulent mixing in the lowest part of the atmosphere. This occurs when the Richardson number is greater than about 0.25 […]*". Following a review of this analysis, consistent with extensive modeling of scenarios typical of LNG facilities, the project team concluded that the transition to a gravity-dominated spread tends to occur within the range of wind speeds already modeled in LNG facility siting scenarios (i.e., at or above 2 m/s). Therefore, defining nil wind based on the critical Richardson number does not introduce 'new' physical behaviors to LNG facility siting studies. Nonetheless, lower wind speeds can have an effect on the spread and mixing of flammable vapor clouds, therefore, another basis for the distinction between low- and nil-wind conditions may be appropriate.

A review of available consequence modeling tools revealed that a commonly used category (integral models) is generally restricted in applicability to wind speeds equal to or greater than 1 m/s, whereas another category (computational fluid dynamics – CFD – models) do not have such limitation. Therefore, a practical definition for nil-wind is "wind speeds less than 1 m/s, measured at a height of 10 m above grade", which leaves "low wind" as "wind speeds equal to or greater than 1 m/s and less than or equal to 2 m/s, measured at a height of 10 m above grade".

## Effects of Nil-Wind

The HSE's review of large vapor cloud incidents found that *"[i]n many of cases examined, 50% (12/24), there is clear evidence from the well-documented transport of vapor in all directions and/or meteorological records that the vapor cloud formed in nil/low-wind conditions. In a further 21% (5/24), the pattern of vapor transport suggests nil/low-wind conditions but there is insufficient data available to be sure*". The authors noted how this frequency is much higher than the overall frequency of nil/low-wind conditions, which they estimated at approximately 5%; consequently, they hypothesized that *"[t]he likely explanation for this finding is that a wider range of smaller losses of containment (with much higher frequency) have the potential to cause a large cloud in these conditions, if the releases are not stopped and the vapor is allowed to accumulate around the source*". However, only a single example (a 2-inch hole in a propane storage vessel, modeled using the DRIFT software) was provided as supporting evidence for this potentially impactful claim.

In order to evaluate the effect of nil- vs. low-wind conditions on the potential hazards of flammable releases, the current study performed a much broader analysis, including numerous scenarios typical of LNG facility siting studies, modeled using the PHMSA-approved CFD tool FLACS. The analysis evaluated both the footprint of the flammable clouds, as well as their VCE potential.

The study evaluated the effects of nil-wind conditions using both a prescriptive and a risk-based approach: the prescriptive approach followed current PHMSA regulations (codified in 49 CFR Part 193) and guidance on LNG facility siting (as provided in https://www.phmsa.dot.gov/pipeline/liquified-natural-gas/lng-plant-requirements-frequently-asked-questions), whereas the risk-based approach applied the quantitative risk assessment (QRA) methodology to LNG facility siting (as specified in NFPA 59A, 2019 Edition). Both approaches used the same generic LNG export facility design as the case study. The facility was assumed to be at early design stage, which allowed a broad set of realistic scenarios to be included in the study, without being affected by site-specific details.

The LNG export facility was assumed to have three pretreatment and three liquefaction trains, each capable of liquefying 5 million tons per year of natural gas from the pipeline utilizing a closed mixed refrigerant loop. The liquefaction process was assumed to use a pre-cooled mixed refrigerant (MR) stream, composed of nitrogen, methane, ethylene, propane, and butane. The refrigerants were assumed to be stored in pressure vessels (bullets) located onsite. LNG was assumed to be stored in two 160,000 $m^3$ atmospheric storage tanks and then pumped to an LNG carrier. The three liquefaction trains and the refrigerant storage area were identified as potential explosion sites (PES), that is, areas where, following the ignition of a flammable cloud, the piping and equipment congestion could potentially lead to damaging vapor cloud explosions. The facility plot plan is shown in Figure 1.

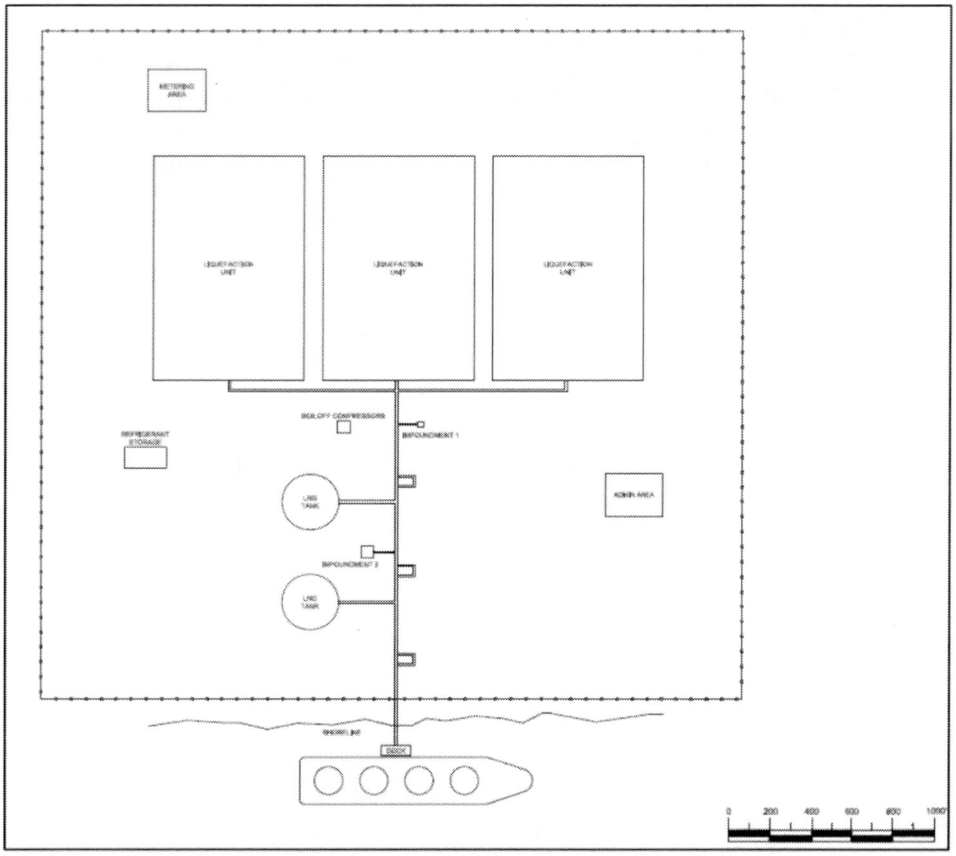

Figure 1. Plot plan for the generic LNG export facility.

## Prescriptive Study

The prescriptive (a.k.a., credible worst case) approach to LNG facility siting specifies the largest release (single accidental leakage source, or SALS) to be considered for each hazardous stream, based on the diameter of the line in question, and requires the hazards from each release to be confined within areas under the control of the facility's owner or transit areas. The scope of the siting study therefore consists of identifying the subset of scenarios and hazards most likely to extend offsite and performing consequence modeling to evaluate compliance with regulatory requirements.

Based on the SALS for the generic LNG facility, the following scenarios were selected for this task to provide a combination of different parameters affecting the behavior of a release (such as momentum, density, location, etc.):

- 2" and 3" holes from the LNG rundown line: these represent a flashing and jetting release of lighter-than-air fluid in proximity of a PES.
- 2" hole in the high-pressure MR liquid line: this represents a flashing and jetting release of heavier-than-air fluid within a PES.
- 1", 2" and 3" holes in the propane storage vessel withdrawal line: these represent a flashing and jetting release of heavier-than-air fluid remote from PES. This release continued beyond the typical 10 minutes, until depletion of the vessel inventory.
- 2" hole in the butane storage vessel withdrawal line: this represents a flashing and jetting release of heavier-than-air fluid remote from the PES and yielding liquid rainout (approximately 23%).
- Guillotine failure of the LNG sendout line: this represents a liquid spill into the LNG impoundment sump, at full sendout rate.

The different hole diameters in two of the scenarios were included to evaluate sensitivity to the release size, particularly to evaluate the accuracy of the HSE report's statement that "*a wider range of smaller losses of containment (with much higher frequency) have the potential to cause a large cloud in [nil/low wind] conditions*". Each of the releases listed above was modeled under four different wind speeds (0, 0.5, 1, and 2 m/s) and for three different wind directions (along, directly opposite, and perpendicular to the release). The equivalent stoichiometric cloud (ESC) volume within each PES, as well as within the entire facility, was calculated throughout each simulation in order to evaluate the effect of nil-wind conditions, as well as other modeling parameters as described above. Figure 2 and Figure 3 show two examples of the ESC volume traces over the entire facility, respectively, for an LNG spill into the impoundment, and a 2" hole in the MR line. In both figures, ESC traces are color-coded by wind speed (red = no wind, green = 0.5 m/s, purple = 1m/s, and black = 2 m/s) and segmented according to wind direction (solid = parallel to release, dashed = opposite, and dotted = perpendicular).

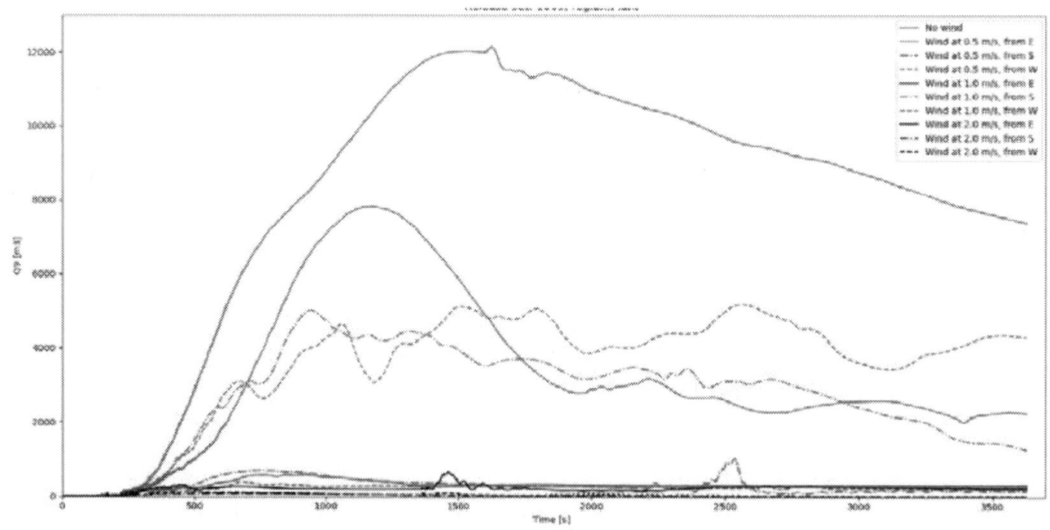


Figure 2. ESC traces for LNG spill into impoundment.

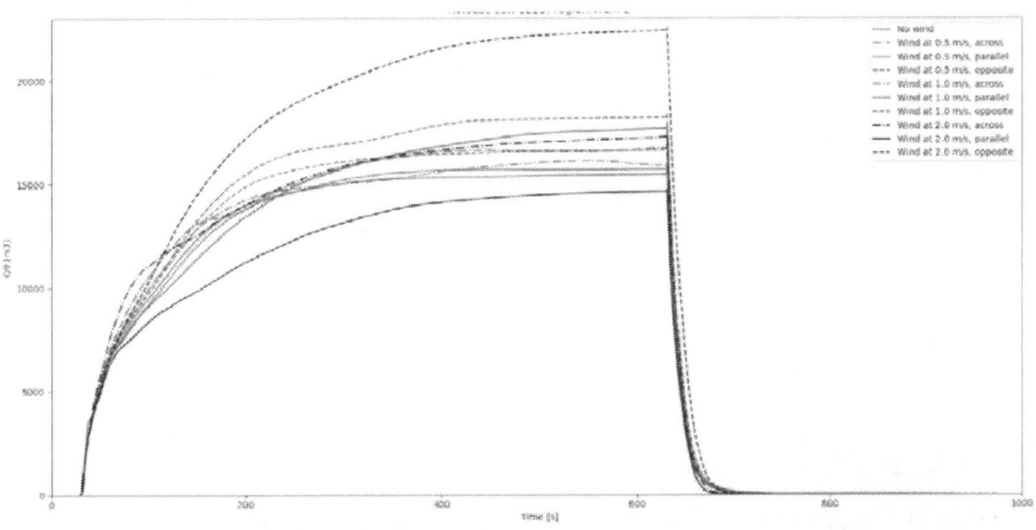


Figure 3. ESC traces for 2-inch MR release.

The results show that nil-wind conditions resulted in a significant increase in the ESC volume over low-wind conditions for the LNG spill. However, the same was not observed for the pressurized MR release, where the difference across wind speeds is fractional; in this case, the largest ESC was observed for 2 m/s wind blowing opposite to the release. Further, in addition to LNG being less reactive than MR, the ESC volume for the LNG spill was also smaller thus would not affect the outcome of a facility siting study.

It should also be noted that these ESC volumes are relative to the entire facility, however, in most cases only flammable cloud volumes within a PES are considered to contribute to overpressure if ignited. The exception to the previous statement is in the event of a deflagration-to-detonation transition (DDT), which requires particular conditions to be present and is beyond the scope of this paper. It is worth noting that the 2017 HSE report postulated the existence of a mechanism known as episodic deflagration, which would induce flame acceleration within very large flammable clouds in the absence of congestion or confinement, thereby causing the entire flammable cloud volume to potentially contribute to a vapor cloud explosion. However, the concept of episodic deflagration has not been generally accepted by the scientific community and has been rejected by several experts ([2], [3], [4], [5]).

Other results from the consequence modeling performed for the prescriptive siting study showed that smaller releases consistently resulted in smaller ESC volumes, thereby contradicting the HSE's argument that small, high frequency releases could form large clouds under nil-wind conditions.

In summary, the results of the prescriptive analysis indicated that nil-wind conditions are unlikely to significantly affect an LNG facility siting study, with the possible exception of uncommon cases where large liquid spills of medium or high reactivity fuels may result in flammable clouds reaching highly congested areas.

## Risk-Based Study

The risk-based approach to LNG facility siting requires every hazardous stream to be considered and, for each stream, various release scenarios need to be evaluated ranging from smaller, more likely releases to larger, less likely releases (including full-bore ruptures). The design for the generic LNG facility was reviewed to identify isolatable inventories, perform parts counts and calculate failure rates for the different representative hole sizes. The consequence modeling, to evaluate both flammable cloud dispersion distances and flammable cloud volumes for overpressures, was performed using FLACS, as in the prescriptive case. However, given the much larger number of scenarios to be modeled (nearly 500 total simulations), the FLACS modeling was limited to releases in the direction of the wind and without any obstacles or obstructions in the 3D model geometry.

The results of the extensive modeling effort are presented in Figure 4 and Figure 5, which present the frequency of cases grouped by percent worsening, respectively, for flammable hazard distance and flammable cloud volume (within the entire computational domain).

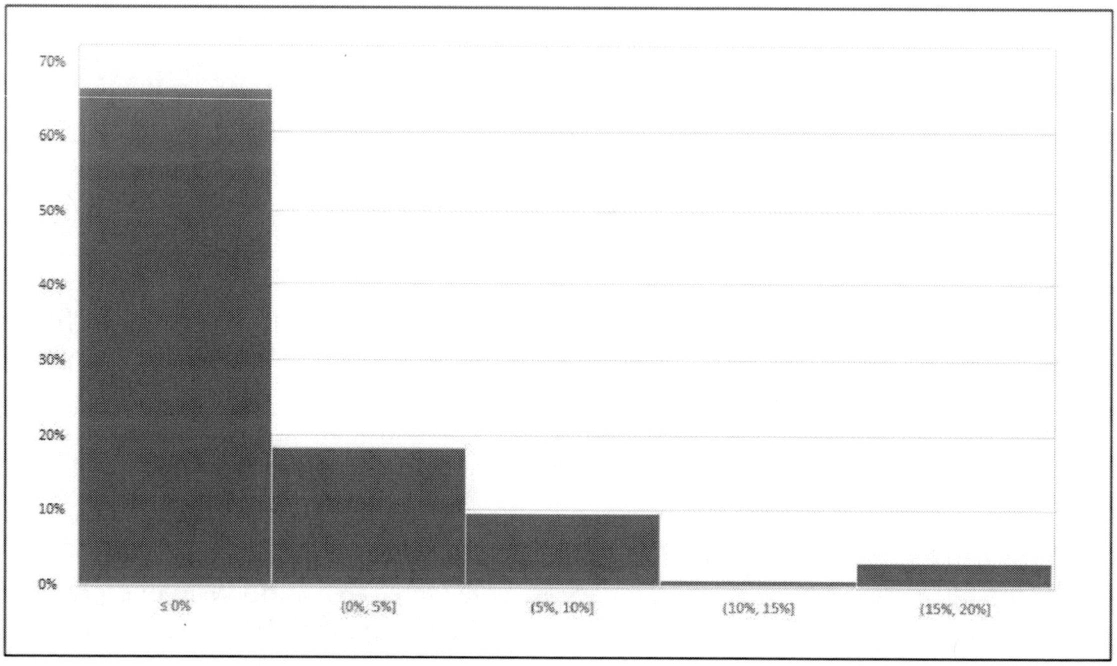


Figure 4: Dispersion modeling results – effect of nil-wind on flammable dispersion distances.

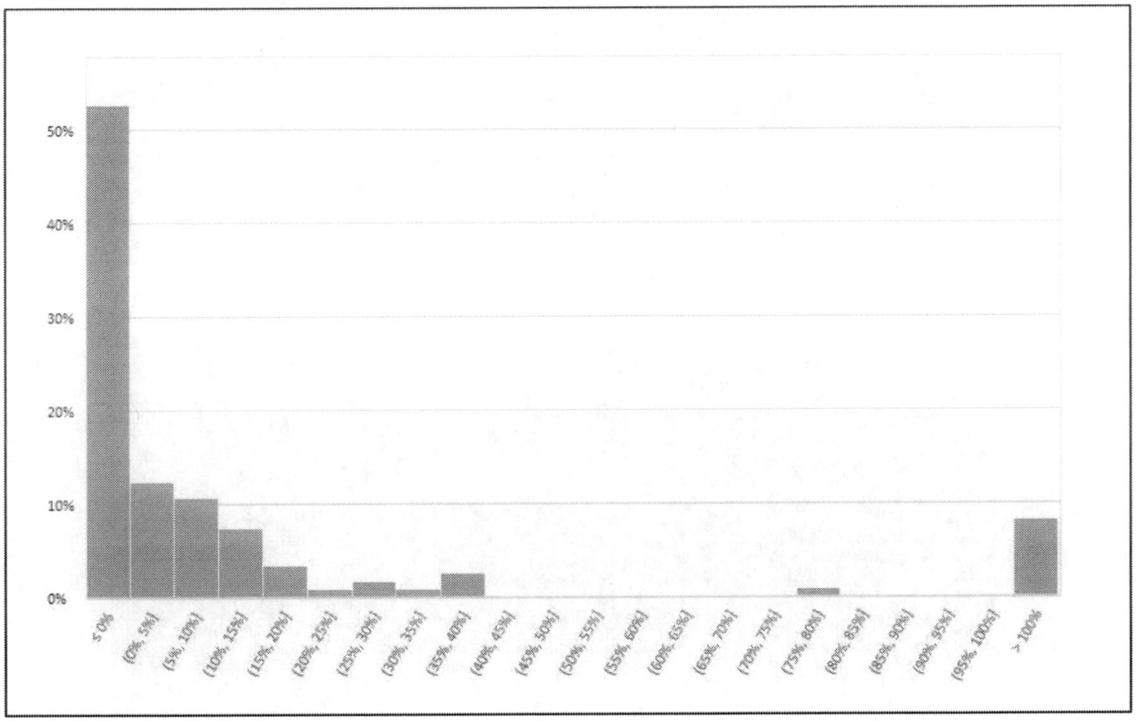


Figure 5: Dispersion modeling results – effect of nil-wind on ESC volume.

The scenarios resulting in the largest hazard increases consist primarily of large-bore releases (6" diameter hole or full line rupture), involving heavy streams (e.g., propane or mixed refrigerant) or low equilibrium pressures (due to flow-limited conditions); several of these scenarios include liquid rainout and pool formation. Pool spreading and vaporization from liquid rainout was conservatively modeled as unconfined by curbs or impoundments, thereby resulting in the largest possible low-momentum vapor clouds. Conversely, the scenarios resulting in the smallest hazard increases (or even in reduction of the hazards) consist primarily of high-pressure releases of lighter streams (LNG or natural gas), where the momentum and turbulence of the release drive the mixing and dilution of the flammable cloud. It should also be noted that the smaller scenarios (up to and in some cases including 2-inch holes) were found to have negligible consequences from a facility siting standpoint (i.e., the distance to LFL was less than approximately 30 ft and/or the maximum ESC volume was less than approximately 1,000 ft3), therefore, the effect of nil-wind conditions on these scenarios was considered irrelevant.

Risk estimation consisted of calculating individual risk of fatality for a hypothetical individual present at a fixed location 100% of the time. Individual risk due to all hazard scenarios was then represented by equal probability of fatality (or iso-risk) contours around the source of risk (i.e., the LNG facility).

The QRA was performed twice – in the first case, following the "traditional" approach which effectively combines nil-wind conditions into the low-wind category (i.e., 2 /s wind speed and F stability), and in the second case the "nil-wind" approach which calculates nil-wind consequences and risk contribution separately from low-wind. The individual risk contours for the two approaches were overlaid as shown in Figure 6, to provide a direct comparison.

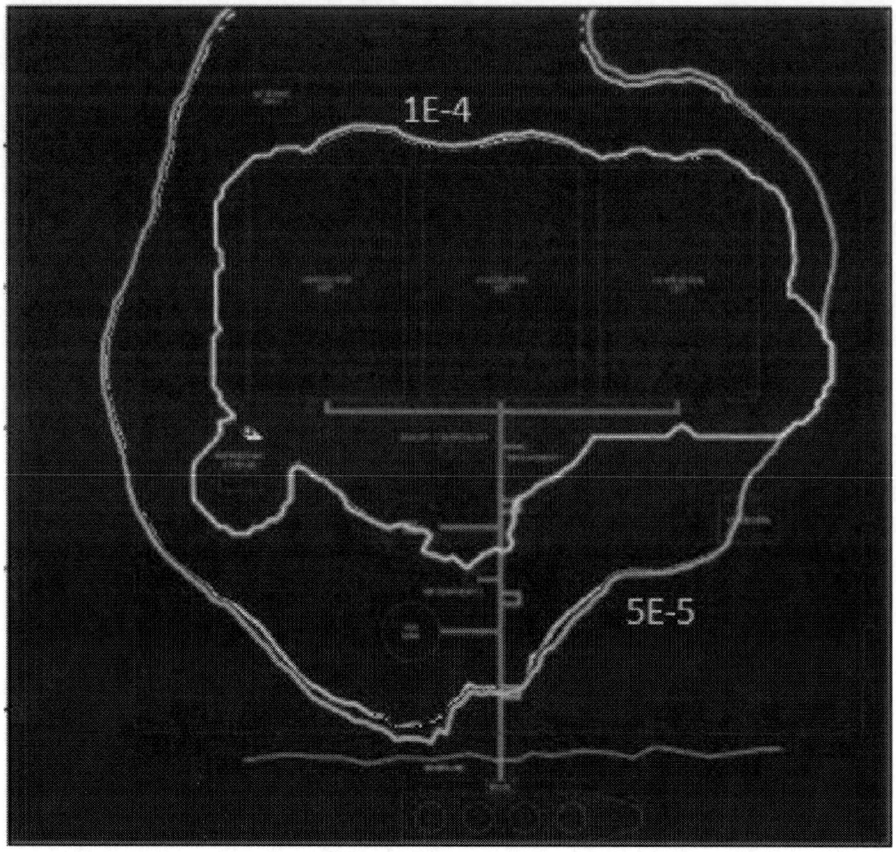


Figure 6: Comparison of individual risk contours: “Traditional” QRA (grey) vs. “Nil-Wind” QRA (yellow).

As shown in Figure 6, the “nil-wind” risk contours are virtually indistinguishable from the “traditional” contours over much of the facility; where the two outcomes deviate, the “nil-wind” contours appear consistently inside the “traditional” ones, indicating a reduction in the overall risk when nil-wind is explicitly factored into the risk calculations.

The QRA comparison performed for this generic LNG facility therefore suggests that the “traditional” QRA approach may be conservative, and that explicitly taking into consideration on a risk basis the effects of nil-wind conditions on hazardous releases may actually reduce the calculated risk. Even though these results are based on a single, generic case study, the effect can be considered broadly applicable: in fact, the consequence modeling results showed that nil-wind conditions can increase the hazard distance primarily for large-bore releases with rainout, whereas they can decrease the hazard distance for smaller bore, high-pressure releases; therefore, considering that small-bore releases are more likely to occur than large-bore ones, it is not surprising that the net effect of considering nil-wind, on a risk basis, may be a reduction rather than an increase of the overall risk.

## Conclusions

Current regulations for siting of LNG facilities specify the range of wind speeds to be considered as those which result in longer predicted downwind dispersion distances than other weather conditions at the site at least 90 percent of the time. Based on these requirements, the bounding cases in most LNG facility siting studies are based on wind speeds of 1 or 2 m/s, depending on the wind data for each site.

Since the consequences of vapor dispersion scenarios, including the size of flammable vapor clouds that may result in explosions, are generally assumed to become worse as the wind speed decreases, the argument has been raised that a worst-case study should consider wind speeds down to zero (i.e., 'nil-wind'). However, before regulations are potentially modified to include broader requirements, the actual effect of wind speed on flammable hazards should be carefully evaluated and include consideration of the frequency of occurrence of nil-wind conditions.

BLUE led a PHMSA-sponsored research project to evaluate the effect of nil-wind conditions on flammable hazards for a wide range of potential scenarios at LNG facilities. The evaluation was performed both under the current prescriptive regulations as well as the risk-based siting methodology outlined in NFPA 59A-2019. The consequence modeling performed under the prescriptive approach, for a set of scenarios typical of LNG facility siting studies, showed that flammable cloud volumes can increase under nil-wind conditions, but primarily for liquid spills or low-momentum releases with significant rainout, whereas pressurized releases are less affected by decreasing wind speed – including several instances in which nil-wind hazards are smaller than under low wind.

The risk-based analysis, which included nearly 500 dispersion simulations performed using a PHMSA-approved CFD tool, confirmed the earlier findings that nil-wind hazards tend to increase primarily for large-bore, low-momentum releases with significant liquid rainout, whereas small-bore, pressurized releases tend to be marginally affected or to result in reduced hazard distances in nil-wind conditions. When the probability of occurrence of different releases is taken into account, the effect on explicitly calculating nil-wind effects appears very limited. In fact, the "traditional" QRA results for the generic LNG facility were shown to be slightly conservative, which can be attributed to the higher frequency of small, pressurized releases for which nil-wind conditions result in equal or smaller hazards.

## References


[1] G. Atkinson, J. Hall, and A. McGillivray, "Review of Vapour Cloud Incidents," Health and Safety Executive, RR1113, 2017.

[2] M. Abdel-jawad and F. Gavelli, "The myth of episodic deflagration in large, unconfined vapor clouds," *Porc. Safety Prog.,* vol. e12162, https://doi.org/10.1002/prs.12162, 2020.

[3] D.M. Johnson and V.H.Y. Tam, "Why DDT is the only way to explain some vapor cloud explosions," *Proc. Safety Prog.,* vol. 36, https://doi.org/10.1002/prs.11874, pp. 292-300, 2017.

[4] G. Chamberlain, E. Oran, and A. Pekalski, "An Analysis of Severe Vapour Cloud Explosions and Detonations in the Process Industries," *Chemical Engineering Transactions,* vol. 77, https://doi.org/10.3303/CET1977143, pp. 853-858, 2019.

[5] S. Davis et. al., "Do not believe the hype: Using case studies and experimental evidence to show why the HSE is wrong about excluding deflagration-to-detonation transitions," *Proc. Safety Prog.,* vol. 38, https://doi.org/10.1002/prs.11998, 2019.

SPRING22
+18TH GCPS
A Joint AIChE and CCPS Meeting

# VCE Potential Cascade Effects

**Cynthia Spitzenberger**
**DNV GL USA, Inc.**
**1400 Ravello Drive, Katy, TX 77449**
**cynthia.spitzenberger@dnv.com**

**Mehmet Ozbey**
**Thornton Tomasetti**
**840 Gessner Rd, Suite 540, Houston, TX 77024**
**MOzbey@ThorntonTomasetti.com**

**Michael Johnson, DNV UK, michael.johnson@dnv.com**
**Gabriele Ferrara, DNV UK, gabriele.ferrara@dnv.com**
**Tatiana Norman, DNV USA, tatiana.norman@dnv.com**
**Stewart Price, DNV USA, stewart.price@dnv.com**

Prepared for Presentation at
American Institute of Chemical Engineers
2022 Spring Meeting and 18th Global Congress on Process Safety
San Antonio, TX
April 10 – April 14, 2022

# VCE Potential Cascade Effects

**Cynthia Spitzenberger**
**DNV GL USA, Inc.**
**1400 Ravello Drive, Katy, TX 77449**
**cynthia.spitzenberger@dnv.com**

**Mehmet Ozbey**
**Thornton Tomasetti**
**840 Gessner Rd, Suite 540, Houston, TX 77024**
**MOzbey@ThorntonTomasetti.com**

**Michael Johnson, DNV UK, michael.johnson@dnv.com**
**Gabriele Ferrara, DNV UK, gabriele.ferrara@dnv.com**
**Tatiana Norman, DNV USA, tatiana.norman@dnv.com**
**Stewart Price, DNV USA, stewart.price@dnv.com**

**Keywords:** Consequence Modeling and Safe Design

## Abstract

The LNG industry has historically maintained a high degree of safety in its operations. This has been built on a clear understanding of the hazards associated with LNG processes, storage and transportation. A vapor cloud explosion (VCE) occurs following the ignition of significant vapor clouds formed by accidental release of flammable fluids and can generate damaging overpressures. LNG, which in its vapor form is natural gas, is inherently safer in this respect compared to other fuels, but this does not remove the need to consider these events.

A key aspect of a VCE is that the resulting damaging overpressures can cause further failures in equipment and as a result, the initial incident escalates. These 'Cascade Effects' can significantly increase the severity and extent of a VCE event. Understanding the potential for Cascade Effects on LNG facilities can inherently improve safety, protecting both people and key energy infrastructure. The US DOT PHMSA commissioned a research project to evaluate the severity of potential explosions and the response of structures to exceptional loads. DNV and Thornton Tomasetti partnered to conduct the research project starting in Q3 2020.

DNV performed dispersion and explosion modeling to evaluate the potential extent of explosion loads from flammable materials present at LNG facilities. Finite Element Analysis (FEA) modeling was performed by Thornton Tomasetti to evaluate the structural response of representative equipment to overpressure loads.

A summary of the research and findings to date are provided within the paper. The research work is ongoing with anticipated conclusion by the end of Q2 2022. Updated findings and conclusions will be provided in future.

*This work was funded in part, under the Department of Transportation, Pipeline and Hazardous Materials Safety Administration. The views and conclusions contained in this document are those of the authors and should not be interpreted as representing the official policies, either expressed or implied, of the Pipeline and Hazardous Materials Safety Administration, the Department of Transportation, or the U.S. Government.*

# 1 Introduction

Cascade effects following a vapor cloud explosion (VCE) have the potential to turn a hazardous release event into an even greater hazardous scenario and loss of containment. Cascade scenarios resulting from the initial scenario may impact emergency responders and complicate their response plan. Greater understanding of the potential for cascade effects and guidance on potential structural impacts will allow the industry to consider these effects and plan for and mitigate their potential.

Understanding cascade potential has two primary aspects, the severity and extent of VCE scenarios that might occur on LNG facilities and the ability of these events to cause damage to structures, equipment and buildings (hereinafter referred to as 'targets'). The physical separation between the explosion source and the target is also a key factor affecting the explosion load. It is essential that guidance is based on a combination of these technical factors with the reality of the actual configurations that occur on LNG facilities.

The overall objective of this research is to provide guidance on the potential cascading effects from flammable vapor cloud explosions and projectiles regarding structural damage at LNG facilities.

In support of the above, the following objectives are outlined:

- Identify the range of LNG facility types and configurations, with an emphasis on defining the explosion sources, flammable inventories and the characteristics of targets.
- Assess the blast capacities of the targets.
- Define the VCE potential of the explosion sources and how this interacts with the targets.
- Develop guidance on the cascade potential that is sensitive to the characteristics of the type of LNG facility.

# 2 Work Plan

To address the research objectives, the project was divided into multiple work tasks as illustrated in Figure 1. The outlined activities are:

1. Review of LNG Facilities
2. Initial Blast Capacity Assessment
3. Screening of Explosion Sources and Targets
4. Detailed CFD Explosion Modeling
5. Detailed Damage Assessment

6. Guidance Development

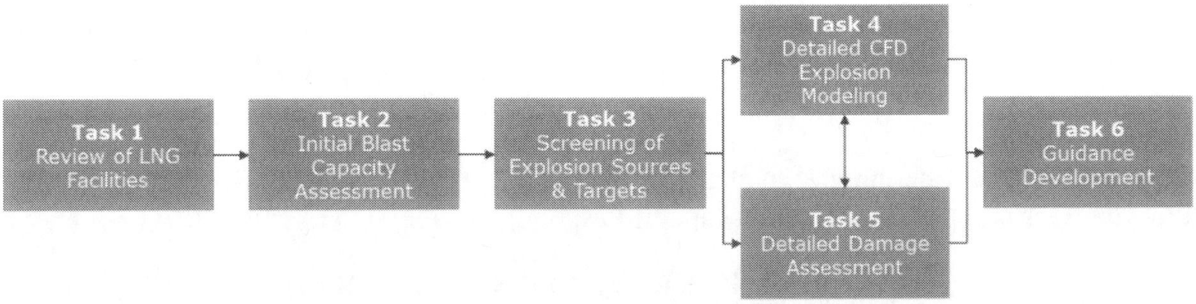


**Figure 1 Research Work Plan Task Overview**

Tasks 1-3 have been completed to date. Task 4 is nearing completion with Task 5 and 6 about to begin. This paper will summarize the outcomes of the completed tasks and provide initial conclusions from the current modeling progress.

# 3 LNG Facilities

The first task was to survey existing and proposed LNG facilities to provide the fundamental data required to ensure that the research is relevant to operating LNG facilities. The output from this task was a description of LNG facilities and equipment, including the following: 1) Capacities, 2) Process (i.e., LNG production, gasification, peak shaving, utilization) 3) Target types (e.g., storage tanks, buildings, road or maritime carriers) with different construction characteristics, 4) Flammable inventories and fuel types, and 5) Explosion sources (typically process congestion and/or confinement).

## *3.1 Types of Facilities*

PHMSA collects and publishes annual data on the number of LNG facilities in the US.[1] As of 2019, there were approximately 217 LNG storage tanks in service with an overall capacity of 56.7 million barrels. These storage facilities vary from marine terminal import/export facilities, peak shaving facilities and satellites to other sites.

In order to establish typical characteristics of the LNG facilities, a survey of existing and proposed installations was conducted. In the survey, publicly available information, from sources such as PHMSA, FERC and the Canadian government, were reviewed and key pieces of information were aggregated.

All types of LNG facilities – existing and proposed; import, export or peak-shaving; small to large – are reviewed for available information. Facilities have different configurations and layouts given their unique siting characteristics and intended purpose. For the evaluation of the cascade effects the intent is to understand the range of types of facilities and the range of distances between hazardous congested areas to major targets. This informs the study of the potential breadth of facility arrangements and equipment that need to be evaluated.

Note that the review was not meant to be exhaustive. Instead, the review was meant to gather enough background material to ensure that the general characteristics of the facility equipment, flammable inventories and equipment arrangements used in the next stages of the research were representative.

In 2019 there were 122 LNG facilities in service and active in the US (excluding mobile and temporary facilities),[1] see Table 1. Base Load facilities are 20% of the number of LNG facilities but represent 59% of the LNG storage capacity with 33% of the number of tanks in operation. Peak Shaving facilities are 57% of the number of facilities and represent 38% of the LNG storage capacity with 47% of the number of tanks in operation. 40% of the Peak Shaver and Satellite facilities have storage with liquefaction capabilities.

**Table 1 US LNG Facility Tank Capacity (2019)**

| Facility Category | Facilities | | LNG Tanks | | LNG capacity | |
|---|---|---|---|---|---|---|
| | Count | % | Count | % | bbls | % |
| Base Load | 24 | 20% | 71 | 33% | 3.37E+07 | 59% |
| Peak Shaving | 69 | 57% | 101 | 47% | 2.17E+07 | 38% |
| Satellite | 20 | 16% | 33 | 15% | 5.15E+05 | 1% |
| Other | 9 | 7% | 12 | 6% | 8.09E+05 | 1% |
| Total | 122 | | 217 | | 5.67E+07 | |

## *3.2 Flammable Inventories*

LNG is natural gas that is cooled to -260°F (-162°C) to become liquid, generally at atmospheric pressure but in some cases at elevated pressures. In the US there are LNG facilities operating in a variety of services. Some facilities export natural gas, some provide natural gas supply, while others are used to store natural gas for periods of peak demand.

The main hydrocarbon flammable materials usually present in an LNG facility are: methane, ethane, ethylene, propane, butane, pentane and hexane. These exist in different quantities and/or as mixtures of several components within the process. The major product streams will differ for the three main types of facilities (peak-shaving plants and LNG export and import terminals), but the main streams are feed gas, LNG, refrigerants, natural gas liquids (NGL) and condensate.

Based on a review of the LNG facilities (peak shaving, import and export terminal) potential release conditions were compiled to represent the major product streams. These representative streams were used in the modeling tasks (3 & 4) to generate release scenarios with the potential to form different flammable clouds and explosion hazards.

## *3.3 Explosion Sources*

Areas of congestion and confinement are located at LNG facilities depending on the amount and type of equipment present for the facility operation and the layout of the facility. If a flammable material is released and forms a gas cloud that overlaps the congested areas and ignites, there is the potential for a vapor cloud explosion (VCE). The level of congestion and size of congested or confined areas vary by type and size of facility. An initial survey of different facilities was conducted to understand the range of congested volumes.

Peak shaving facilities operate at a wide range of capacities (1-750 MMCF/D maximum vaporization capacity; 400 to over 1.16 million bbl tank volume). Smaller facilities will have congested volumes in the range of 150-200 $m^3$. Larger facilities may have estimated congested volumes up to 9,300 $m^3$.

For import / export LNG facilities in the US, depending on the facility layout there could be numerous regions of congestion and confinement that could lead to an explosion hazard. The liquefaction or vaporization equipment areas are normally the largest sources of congestion on the facility. If congested regions are close together without adequate separation distances, they may combine and act as one large, congested region in a deflagration incident. The reviewed import facilities have congested volumes in the range of 375-70,000 $m^3$. The congested volumes for export facilities vary greatly. Some of the facilities surveyed have a congested volume greater than 30,000 $m^3$; while others range from 8,000-13,000 $m^3$.

### *3.4 Target Equipment Types*

Key targets for evaluation of cascading effects are equipment or structures that will lead to further loss of containment of hazardous material. Such items are mainly the large storage equipment present at the site. Large storage vessels include the following:

1. LNG storage tank(s)
2. Refrigerant storage tank(s)
3. Condensate storage tank(s)
4. Other large process vessels
5. Large piping and piping supports
6. Maritime carrier
7. Road tanker

The review of different equipment types and configurations present across the range of LNG facility types indicated that 16 different representative equipment target types would need to be evaluated. For the detailed finite element analysis (FEA) modeling, structural details were required as input, such as materials of construction, thicknesses, support structures, etc. These details were gathered for each of the representative equipment types.

The following sub-sections characterize a few of the main equipment types that were selected for the evaluation.

#### 3.4.1 LNG storage tanks

LNG storage tanks usually represent the largest onsite inventory at the facilities and thus are a key concern with relation to cascading effects. There are four types of LNG storage tanks as defined in NFPA 59A[2]:

1. *Single-containment tank system*: "system in which the self-standing inner (i.e., primary) container is designed to contain LNG and is surrounded by a separate container that is not designed to contain LNG." (NFPA 59A,[2] 3.3.5.4.4)
2. *Double-containment tank system*: "a single-containment tank system surrounded by and within 20 ft (6 m) of a self-standing containment wall (i.e., secondary container) that is open to the atmosphere and designed to contain LNG in the event of a spill from the primary or inner container." (NFPA 59A,[2] 3.3.5.4.1)
3. *Full-containment tank system*: "a tank system in which the self-standing inner (i.e., primary) container is surrounded by a separate self-standing secondary container designed to contain LNG in the event of a spill from the inner container, and the secondary container is enclosed by a steel or concrete roof designed such that excess vapor caused by a spill of LNG from the primary container will discharge through the relief valves." (NFPA 59A[2] 3.3.5.4.2)
4. *Membrane-containment tank system*: "a tank system consisting of a thin metal liquid barrier and load-bearing thermal insulation supported by a self-standing outer container jointly forming an integrated composite tank system designed to contain liquid and vapor during tank operation as well as LNG in the event of leakage from the liquid barrier, and where the vapor-containing roof of the outer container is either steel or concrete configured such that the excess vapor caused by a spill of LNG from the liquid barrier will discharge through the relief valves." (NFPA 59A,[2] 3.3.5.4.3)

API 625[3] provides additional definition of the different types of tank systems and design requirements. Older facilities primarily used single or double wall containment LNG tank systems. A survey of the type of LNG storage tanks worldwide in 2003 indicated that over half were double wall steel tanks.[4-5] More recent LNG facility design permits / proposals tend to involve full-containment tank systems. As the LNG storage tanks are often the greatest inventory of concern for cascade effects, these tank structures will be a focus for the detailed structure modeling.

### 3.4.2 Refrigerant storage tanks

Propane and Ethylene are materials commonly used as refrigerants in LNG facilities. Aside from the LNG, these tanks are often the next most significant flammable inventories stored onsite. Propane may be stored in spherical tanks, horizontal bullets or vertical bullets. Propane is stored in ambient temperature, pressurized, single wall tanks designed to ASME BPVC (Boiler and Pressure Vessel Code) Section VIII Division 1.[6] Ethylene is normally stored in pressurized, double wall vacuum jacketed horizontal vessels. The vessels could also be oriented vertically. The ethylene vessels would also be designed to ASME BPVC Section VIII Division 1.[6]

### 3.4.3 Large piping and piping supports

Piping and pipe racks are usually present at the facility to transfer inventories and feed components to the process. Depending on the size and layout of the facility, long pipe runs may be required to transfer LNG to/from the storage and liquefaction / vaporization trains. Long pipe runs may also be required for the transfer of LNG to / from the marine carrier at import / export facilities.

Refrigerant storage must also be supplied to the process trains via piping. Large diameter pipe in a long pipe run can contain a significant amount of inventory that could potentially be released.

Usually, the long transport lines on the facilities have significant wall thickness and can withstand some high pressure loads in a VCE. However, the supports to the pipelines can sometimes be the weak point and fail when exposed to high loads.

### 3.4.4 Buildings

Building structures may also be impacted from an explosion event and in relation to cascade effects the damage could impact process safety control equipment or hazardous equipment contained within a building. Building structures and designs vary depending on when constructed and site requirements. Several standards and guidelines already exist regarding the siting of occupied buildings in relation to explosion hazards, such as API 752 and API 753.[7-8] Typical building examples are used for the load analysis to generalize the potential damage.

# 4 Initial Blast Capacity Assessment

Thornton Tomasetti (TT) generated structural analytical models of the selected targets of interest. More specifically, TT used Single-Degree-of-Freedom (SDOF) analysis and Finite Element (FE) models in Abaqus with geometric nonlinearity (i.e., capturing tension membrane, second order effects) to determine the structural response to the explosion. The explicit dynamic solver in Abaqus is proven to be capable of efficiently capturing impulsive and impact loading responses having large material and geometric non-linearities.

Each reported target blast capacity was based on any damage mechanism identified that can cause leakage and lead to cascading effects. Target blast capacities are provided in terms of peak blast pressure and duration assuming a shock loading consistent with a detonation. A shock loading was selected because pressure waves propagating away from a VCE will 'shock up' even if not initially a shock wave and the most severe events, where cascade effects are more likely, will originate as shock waves. The effect of non-shock loading will be assessed in more detailed analysis in Task 5.

FE models of LNG storage tanks included the static fluid pressure. Constitutive material models included strain-rate dependent elastic-plastic material effects. An element deletion technique was used based on the material failure criteria based on strain levels. The mesh density used was sufficient to capture the local buckling modes. The beneficial effects of modeling the vapor space pressure increase during the roof response was also included.

All FE models utilized plastic strain limit criteria to achieve more realistic and less conservative results. Detailed information on the methodologies used for the FE and SDOF models can be found in the later sub-sections.

Performance of the target structure under the prescribed blast loads was based on the evaluation of the structural components under flexure and overall frame response for lateral loading. The secondary framing and cladding components were analyzed using the nonlinear SDOF methodology, which is widely recognized and accepted as an industry standard within the blast community. When deemed necessary, typically for assessing the global response of a structure or where multiple structural element types and load conditions are used, the structural components for the primary structural framing were analyzed utilizing a nonlinear dynamic finite element model (FEM). In these specific cases, such as tapered moment resisting frames, the SDOF methodology would not be able to capture the full interaction between structural members and can significantly under-predict the blast response.

### *4.1 Finite Element (FE) Analysis Methodology*

In a Finite Element (FE) analysis, the tank, piping, or vessel is modeled almost in its entirety. In this type of analysis, the material response in terms of displacement, stress and strain can be calculated at multiple locations thus resulting in a more comprehensive analysis. The present study is based on the Ultimate Strength Analysis (sometimes called Ductility Level Method) approach. The Ultimate Strength Analysis is the most refined analysis type listed in API RP 2FB[9] and presents the lower level of conservatism. For the purpose of this project, ultimate strength analysis against blast loading was conducted using a general-purpose finite element analysis package, Abaqus.[10] Abaqus uses an explicit integration scheme which is well suited for performing dynamic, transient analysis of problems where large displacements take place by taking plasticity and failure into account.

In nonlinear problems – those with large deformations and/or nonlinear material response – the highest frequency of the model will continually change, which consequently changes the stability limit. Abaqus has fully automatic time incrementation strategy (where the code accounts for changes in the stability limit) for time incrementation control.

### *4.2 Single Degree of Freedom (SDOF) Analysis Methodology*

A nonlinear dynamic SDOF approach was used to analyze the building structural components members as a part of this task. In conventional structural design, the minimum mechanical properties of steel materials (e.g., yield strength and tensile strength) are used in the analysis. In practice, the average yield strength of steel materials is approximately 10% greater than the specified minimum values. Thus, a Strength Increase Factor (SIF) is typically used to account for this condition in blast analysis.

Additionally, under rapidly applied loads, such as blast loading, steel materials experience an increase in strength. The magnitude of dynamic increase is dependent upon several factors, including static material strength and strain rate. In blast analysis, a Dynamic Increase Factor (DIF) is applied to static strength values to incorporate the effect of material strength increase with strain rate.

The SIF and DIF values used by TT are those suggested in "American Society of Civil Engineers' Design of Blast Resistant Buildings in Petrochemical Facilities".[11]

### *4.3 Blast Capacity Results*

The blast capacity overview of each equipment type is summarized in Table 2. In the table, cascade failure description, overpressure and its duration are included. Please note that the blast capacity was estimated in the condition that anchor bolt failure has not occurred, which can potentially be the limiting component for tall tanks anchored to concrete foundations.

The results of this study show that most typical equipment that can lead to cascading effects at an LNG facility have significant inherent blast resisting capacity. Refrigerant storage and other large pressure vessels require loads greater than 50 psi (3.45 bar), 100 ms duration to result in overturning failures (saddle failure), which is very hard (if not impossible) to achieve from typical vapor cloud explosions seen in the industry.

Lowest blast resisting capacity was observed for pipe supports, although piping themselves have significantly higher blast resisting capacities. Among all storage tanks, single containment tanks have the lowest blast resisting capacity. On the positive side, single containment tanks have considerable blast resisting capacities that can only be exceeded by large events at very close distances. Hence, cascading effects can be avoided by providing proper separation distance from the identified explosion sources.

Refinements of the finite element models will be performed as needed based on the potential blast loads on each target calculated in Task 4. Additional internal content fill conditions, anchoring effects including anchor bolts, and nozzle to pipe connections will be considered in Task 5, once magnitude of potential blast loads are more refined as a result of Task 4.

**Table 2 Summary of Blast Capacities (Applied Overpressure)**

| Equipment | Failure Description | Overpressure (psi) , Duration(ms) [0.0, 4.0, 8.0, 12.0, 16.0, 20.0, 24.0, 28.0, 32.0, 36.0, 40.0, 44.0, 48.0, 52.0, 56.0, 60.0, 64.0, 68.0, 72.0, 76.0, 80.0, 84.0, 88.0, 92.0, 96.0, <100] |
|---|---|---|
| Storage Tanks (Single Containment) | Inner tank roof weld line failure | 6 psi, 100ms; 10 psi, 50ms; 22 psi, 25ms |
| Propane Storage Tank (Single Shell Horizontal) | Overturning failure (shell thick. 2.25") | 60 psi, 100ms; 70 psi, 50ms; 90 psi, 25ms |
| | Tank burst (shell thick.1") | 26 psi, 100ms |
| Ethylene Storage Tanks (Double Shell Horizontal) | Overturning failure | 52 psi, 100ms; 70 psi, 50ms; 95 psi, 25ms |
| Other Large Process Vessels (Vertical Bullet Tanks) | Overturning failure 1" thick. | 55 psi, 100ms |
| | Overturning failure 2" thick. | 95 psi, 100ms |
| | Overturning failure 6.69 thick. | 180 psi, 100ms |
| Large Piping and Piping Supports | Pipe support guide failure | 2.5 psi, 100ms; 5 psi, 50ms; 10 psi, 25ms |
| | Pipe support frame failure | 13.2 psi, 100ms; 25 psi, 50ms; 37.5 psi, 25ms |
| | Flange bolt failure | 16.1 psi, 100ms; 27.5 psi, 50ms; 40.0 psi, 25ms |
| Maritime Carrier | Inner tank failure | 80 psi, 100ms; 100 psi, 50ms; 145 psi, 25ms |
| Road Tanker | Overturning | 15 psi, 100ms; 20 psi, 50ms; 25 psi, 25ms |
| | Tank burst | 30 psi, 100ms; 30 psi, 50ms; 30 psi, 25ms |

# 5 Screening of Explosion Sources and Scenarios

The purpose of this activity was to conduct an initial screening exercise to reduce the number of targets and explosion scenarios that need to be analyzed in the later detailed modeling activities (Tasks 4 and 5). Based on a combination of the output from Tasks 1 and 2, simple and conservative source explosion and vapor dispersion modeling was carried out to determine the size and severity of VCE scenarios based on key defining event characteristics. The output from Task 2 (Initial Blast Capacity Assessment) provided an indication of the criteria loads required to cause damage to a range of targets.

## *5.1 Congested Regions*

Based on the review of selected representative peak shaver and terminal facilities, the range of possible congested regions at the various sites is demonstrated. There is not a clear correlation between facility capacity and the potential size and overall volume of congested regions at a facility. As expected, the congested regions are dependent on facility specific design factors that are often unique, such as the facility property area and layout design. At a particular facility, there can be a range in size of congested regions from small to large. Generally, the average congested volume of a region on the peak shavers is smaller than the average congested region on a terminal facility; this is also as expected given that generally the terminal facilities are larger sites than the peak shavers.

The conclusions regarding the congested regions are coarse, but at least indicate a likely range of congested regions that may be present at the different facility types. The minimum and maximum ranges indicated by the review are 97 $m^3$ to 9,300 $m^3$ at peak shaver facilities and 375 $m^3$ to 70,000 $m^3$ at terminal facilities. Note that the estimated congested regions are based on aerial layout surveys of the facilities and not site visits; heights were estimated from topography data for nearby structures or assumed. Also, the estimated congested volumes are the entire region volume with no deductions for space occupied by equipment.

There is also wide variation on the level of confinement for the congested regions across the facilities. Some facilities have little to no areas with confinement; while others have significant areas of confinement with overhead equipment or weather shelters for the equipment.

## *5.2 Flammable Dispersion*

The Task 1 activity compiled a list of potential release scenarios of different hydrocarbons that may be present at LNG facilities. Not all release scenarios would be applicable to all facilities, as not all facilities have refrigeration capabilities and there are different LNG technologies and designs.

For Task 3, a selection of representative scenarios was modeled with DNV's Phast v8.4 software package to estimate the potential cloud sizes that may develop for the various scenarios. 2 inch and 6 inch releases were modeled to be representative for potential release scenarios at the facilities, consistent with NFPA 59A. The releases were assumed to reach steady state conditions.

The largest flammable stoichiometric cloud volumes were formed from releases with large hole sizes in calm ambient conditions. A majority (76%) of the 2 in release cases generate flammable stoichiometric cloud volumes less than 5,000 $m^3$. Only four of the 33 modeled release cases generate cloud volumes greater than 10,000 $m^3$ with the greatest stoichiometric gas cloud volume of ~40,000 $m^3$.

The 6 in release cases generate larger cloud volumes, with the largest volume of ~319,000 $m^3$. A majority (67%) of the 6 in release cases generate stoichiometric cloud volumes greater than 50,000 $m^3$. Note the generated hazard zones should be considered conservative in that no safety measures are accounted for in the estimate of the cloud volume, such as trenches or bunding that would limit the spread of the LNG release and thus the vaporization and gas cloud formation.

These cloud volumes demonstrate that it would be possible to have releases that could engulf the larger congested regions at the LNG facilities. However, other factors such as location of inventories in relation to congested regions would apply. The flammable cloud heights remain relatively low, under 15 m, for all scenarios, indicating that the clouds have a wide, flat shape.

### *5.3 Peak Overpressures*

Leaks from pressurized gas pipework or equipment have the potential to form a flammable gas-air mixture under certain conditions around the leak location.

- If a flammable gas air mixture is ignited in an open area it would burn without generating a damaging blast wave. This type of "flash fire" event would endanger personnel within the flammable cloud but not cause significant damage to equipment or personnel outside the cloud envelope.
- If the flammable gas air mixture is ignited in a region of the facility that is filled with obstacles (such as pipework) or in an enclosure (such as a compressor house) there is the potential for a blast wave to be generated that could have a damaging impact outside the immediate flammable area.

An understanding of the mechanisms required to cause damaging levels of pressure to be generated is needed to identify situations where this explosion potential is of concern.

#### 5.3.1 Multi-Energy Method Model

The Multi-Energy Method (MEM) detailed by TNO[12] and GAMES projects[13] was applied as the initial method for calculation of the peak overpressure from a defined congested region. Application of the model can be performed simply and quickly to estimate the overpressure based on a congested region's characteristics.

The MEM model requires definition of key parameters related to the congested region.

- *Volume Blockage Ratio (VBR)* is defined as the ratio of the obstacle volume inside an obstructed region over the volume of the obstructed region. The obstructed region is the potential source for the overpressure generation in a vapor cloud explosion. The TNO Yellow Book[14] outlines a methodology for determination of the obstructed region volume.

The method consists of defining an initial obstructed region usually in the basic shape of a rectangular box, and with consistent refinement of the volume space to remove empty space of the box that does not contain congestion. The obstacle volume is meant to represent the volume in the region occupied by the various vessels and equipment that the flammable vapor cloud would not be able to occupy and disperse through.

- *Mean Diameter (D)* is meant to provide an average value for the entire obstructed region, assuming a homogeneous distribution of obstacle types and obstacle diameters. It is meant to represent the average obstacle size in the obstructed region, based on diameter. This parameter may be defined via several different approaches, such as with arithmetic mean, harmonic mean or hydraulic diameter. For purposes of this study the arithmetic mean is assumed applicable.
- *Flame path length* is the distance from the point of ignition to the edge of the cloud or the edge of the congested region. Often for simplicity, the point of ignition is assumed at the center of the obstructed region and then the flame path length would be half of either the length, width or height dimensions of the region. If specific ignition locations are considered, then the flame path length may be calculated based on the longest potential distance of the burning cloud within the obstructed region.

### 5.3.2 MEM Calculations

Based on an understanding of the range of the congested region volumes that would be present at LNG facilities, the MEM calculations were used to estimate the potential peak explosion overpressures based on a representative range of factors. The following ranges were applied for the calculations:

- Volume: 10 to 50,000 $m^3$ assumed as hemisphere volumes to represent the potential facility congested regions. The length of the flame path is calculated as the radius of the hemisphere.
- VBR: based on DNV's experience the VBR is often less than 30%. Three representative values – 5%, 10% and 20% – were assumed.
- Diameter (D): based on DNV's experience three representative D values were assumed – 0.05 m, 0.15 m, and 0.30 m. These D values relate approximately to equipment diameters of 2, 6 and 12 in. Note that the values for D have been chosen to demonstrate the potential range and in order to be conservative at this stage.

Based on the assumed key parameters the peak explosion overpressure was calculated based on the 3D and 2D expansion equations. Results are presented for methane and ethylene to demonstrate a low and high reactivity materials (differences in burning velocity), respectively. Ethylene is not always present at LNG facilities; only some facilities use it as a refrigerant component.

Note that the graphs are limited with an upper limit of 10 bar (145 psi) overpressure as normally the transition from deflagration to detonation (DDT) can occur for hydrocarbons other than methane once pressures exceed 1 bar (14.5 psi) and the MEM correlation is only representative of deflagrations within a congested region.

As shown, in the figures, as the region volume increases, the predicted peak overpressure increases, if all other variables are constant. This results because the flame path term is assumed to be the radius of the hemisphere volume; so the greater the volume the longer the flame path which means greater opportunity for the flame to accelerate as it travels through the congested region.

The curves resulting in the highest overpressure, have high VBR and low D values. The higher levels of congestion associated with high VBR values are key in fostering the development of high explosion overpressures. D values have the inverse relationship. Smaller diameter equipment facilitates the flame acceleration in the burning cloud front. As the flames accelerate across the region, higher overpressures are generated.

The range of parameters are chosen to demonstrate conservatively the potential pressure magnitudes. Some combinations such as large volumes of 0.05m D and 20% VBR would not be likely given the high level of congestion this would represent and is assumed not typical at facilities, especially for large volumes.

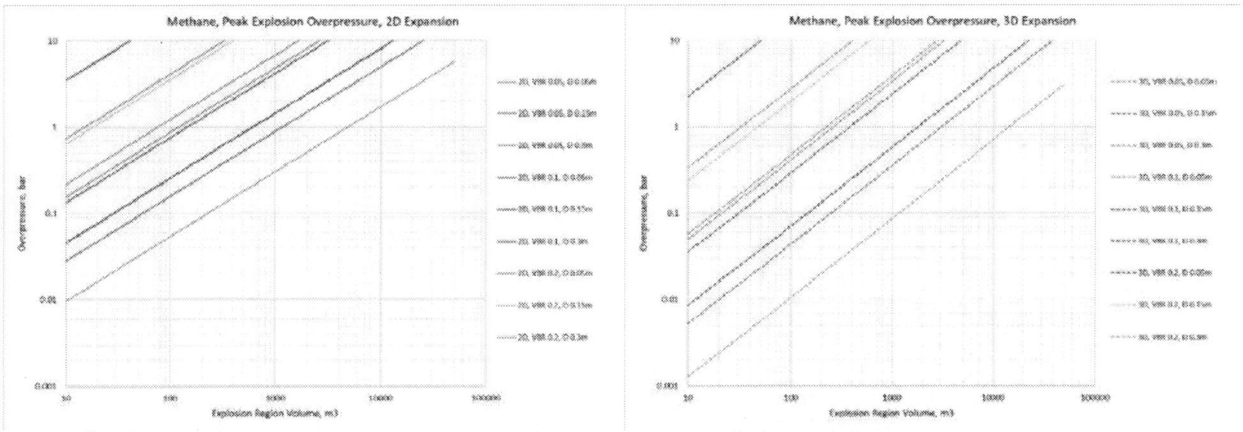


**Figure 2 Peak Overpressure Estimate, Methane**

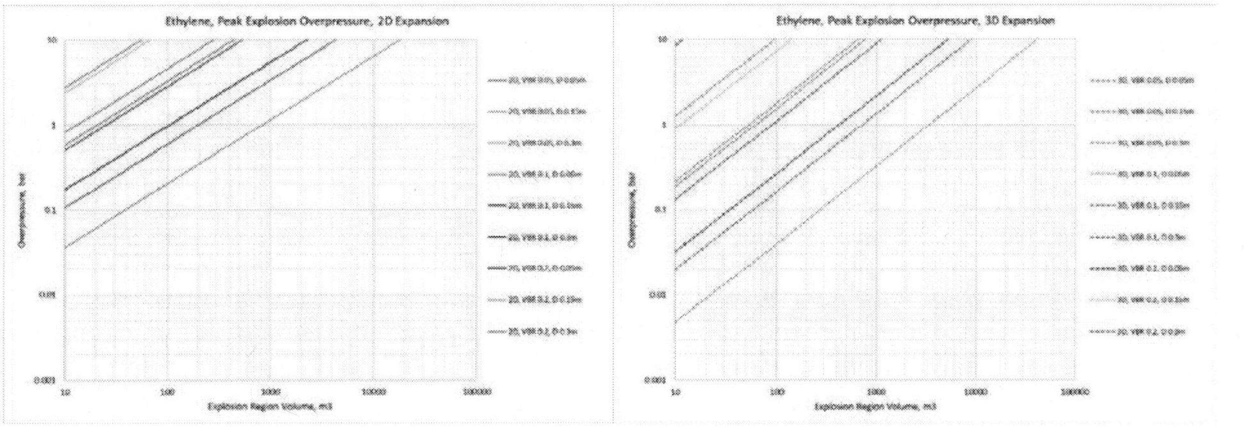


**Figure 3 Peak Overpressure Estimate, Ethylene**

#### 5.3.3 Pressure Decay

Outside of the explosion source region the overpressure will decay with distance. The pressure decay varies on the size of the source explosion. The peak overpressure is sustained out to 5 m for 100 $m^3$, 10 m for 1,000 $m^3$, and 20-30 m for 10,000 $m^3$, depending on the multi energy curve level / originating peak overpressure. Overpressures are less than 0.2 bar (2.9 psi) for all curve levels by 30 m, 70 m, and >100 m for the respective congested regions. By a distance of 100 m, the overpressure decreases by a minimum of 91% for 100 $m^3$, 80% for 1,000 $m^3$, and 53% for 10,000 $m^3$.

#### 5.3.4 Detonation

The scientific community has recently suggested that detonation events may have characterized major VCE accidents that have occurred in the industry with higher incidence than assumed in the past.[15] DNV has conducted a substantial amount of large and full-scale experimental research into VCEs over a period of four decades. Based on this research, we consider that DDT can be excluded from VCEs involving LNG vapor alone. Though DDT is technically feasible for high methane (>90%) vapors, the difficulties in achieving this means that such an event is not credible for real facilities. DDT will therefore only be considered for higher hydrocarbon releases, such as those that might occur following a release of refrigerant. This has the consequence that sites that are limited to just LNG inventory, such as gasification facilities, will only have VCEs that involve deflagration produced by, and limited to the extent of, the congested or confined region the vapor cloud is in or engulfs. Pressure generation will not extend beyond this into any part of the vapor cloud outside the region. Again, of note, methane deflagrations can still be severe, producing supersonic flames, as was found by DNV experimental studies in the 1980's[16] and more recently in experimental studies at SRI International.[17]

DDT will occur at some point between 1 and 10 bar (14.5 – 145 psi), depending on the scenario. Where DDT is considered, the decay of pressure from the cloud boundary is a special case. These events are generally at the top end of the VCE damage potential for items within the detonating cloud, but for large 'pancake' clouds the pressure magnitude decays rapidly from the cloud edge (though not the impulse). The DDT research identified that for these large clouds, pressure magnitude was related to only the height of the cloud and the distance from the cloud edge and had no dependence on the overall cloud volume.[18-19]

### *5.4 Comparison to Damage Thresholds*

The overpressure curves are compared to the damage thresholds predicted from Task 2. Figure 4 compares the 2D expansion overpressure curves for methane to the lowest overpressure value potentially capable of causing equipment failure with the potential of causing cascade effects.

Note that the MEM peak overpressure estimate is representative of the potential side-on overpressure within the congested region; outside of the congested region, the overpressure would decay from the peak value with distance from the region. Equipment items within the congested region may be exposed to the peak overpressure. If the equipment item is outside of the explosion region, then in a deflagration scenario the equipment exposure would likely be a decayed overpressure less than the peak pressure. The MEM calculations represent an idealized situation;

in reality there may be geometry arrangements and confinement that could generate greater overpressure values in localized areas than predicted.

As shown in Figure 4 the worst combination of D (0.05 m) and VBR (20%) generates peak overpressures greater than the equipment thresholds, for almost all volumes. For the worst-case D (0.05 m) with VBR greater than 10%, the results indicate that congested region volumes greater than 200 $m^3$ generate overpressures exceeding all the considered equipment thresholds. For the worst-case VBR (20%) with D values of 0.3 m or less, congested region volumes greater than 1,000 $m^3$ generate overpressures exceeding all the considered equipment thresholds.

Some target categories, e.g., Large Piping and Piping Supports, have low thresholds (2.5 psi, 0.17 bar) that can be damaged by the majority of the representative congested region cases shown. Additionally, once the volume reaches 50,000 $m^3$, even the weakest combination of D-VBR values is capable of surpassing the minimum threshold of all the target categories.

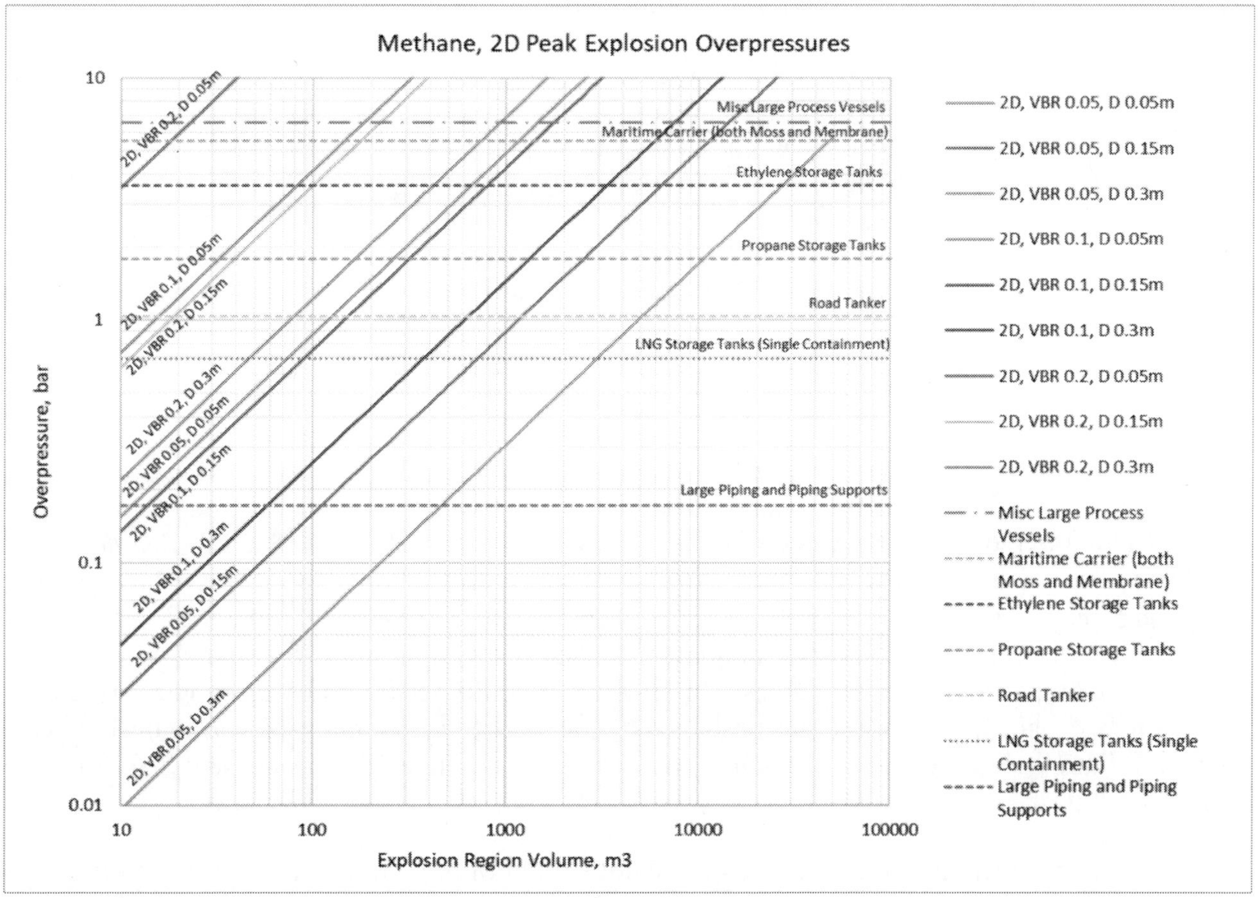


**Figure 4 Peak Overpressure and Damage Thresholds Comparison, 2D, Methane**

When comparing the overpressure damage thresholds to the MEM overpressure calculations, it is shown which combination of congestion characteristics, confinement, material, and volume lead to peak overpressures that can exceed the minimum damage thresholds for the different equipment groups. The following explosion regions were noted for further evaluation in the next project phase as a minimum.

- VBR 20%, range of D values (0.05, 0.1, 0.3 m), congested volumes of 100; 500; 1,000; 1,500 $m^3$
- VBR 10%, range of D values (0.05, 0.1, 0.3 m), congested volumes of 100; 350; 1,000; 5,000 $m^3$
- VBR 5%, range of D values (0.05, 0.1, 0.3 m), congested volumes of 200; 1,000; 10,000 $m^3$

## 6 Current Work

The previous sections have summarized the work completed to date for Tasks 1-3 of the research project. Task 4, Detailed CFD Explosion modeling, is currently in progress with completion expected in Q1 2022. Dispersion and explosion modelling has been performed for a range of scenarios in terms of the release conditions and the explosion regions. The FLACS CFD code is used for this modeling. In addition to providing explosion predictions for deflagration events, the FLACS explosion predictions are used for higher hydrocarbon (C2+) VCEs to indicate when DDT is possible, and the detonation of the full cloud should be considered.

The output from Task 4 will be input for Task 5 (Detailed Damage Assessment), comprising detailed pressure loading data for targets and missile characteristics and energies. TT will use the applied blast load time histories generated from the Task 4 CFD calculations to recalculate the damage to the FE models of the targets of interest in greater detail.

The final activity, Task 6 is the Guidance Development which will build upon the research findings to offer guidance for site preparation for and prevention of cascade effects related to VCE hazards.

## 7 Conclusion

The initial research work has provided a summary of the range of LNG facilities operating in the US. Although there are different types of facilities, with different operations and capacities, each has inventories of flammable hydrocarbons and areas of congestion that could potentially result in a vapor cloud explosion. The range of possible equipment types were defined for evaluation for the blast failure analysis. The results of the blast capacity study show that most typical equipment that can lead to cascading effects at an LNG facility have significant inherent blast resisting capacity which is a credit to design standards. Initial dispersion and MEM explosion modeling indicates the types and sizes of congested regions that could generate deflagration overpressures in the range of the damage thresholds predicted by the blast capacity analysis, within the congested region. Equipment separated a distance outside of the congested region will have decayed explosion loads.

The current and next phases of the research plan to perform more detailed modeling to refine the overpressure and equipment failure damage estimates. Based on the conclusions from the detailed modeling, future guidance would hope to provide advice regarding separation distances, deflagration versus detonation differences, volume congestion and confinement and the relation to cascade effects.

# 8 References

[1] US DOT PHMSA. *Gas Distribution, Gas Gathering, Gas Transmission, Hazardous Liquids, Liquefied Natural Gas (LNG), and Underground Natural Gas Storage (UNGS) Annual Report Data*. https://www.phmsa.dot.gov/data-and-statistics/pipeline/gas-distribution-gas-gathering-gas-transmission-hazardous-liquids . Accessed 12 January 2020.

[2] National Fire Protection Association (NFPA). *59A Standard for the Production, Storage, and Handling of Liquefied Natural Gas (LNG)*. 2019. NFPA 59A-2019.

[3] American Petroleum Institute. *Tank Systems for Refrigerated Liquefied Gas Storage.* August 2010, Addendum 3 - June 2018. API 625, 1st Edition.

[4] Woodward, John L. and Pitblado, Robin M. *LNG Risk Based Safety*. s.l. : John Wiley & Sons, Inc., 2010. ISBN 978-0-470-31764-8.

[5] Coers, DH. LNG storage, fundamentals of base load LNG. *Gas Technology Institute Training Course.* Linthicum Heights, MD : s.n., Sept. 22-25, 2003.

[6] ASME. *Boiler and Pressure Vessel Code, Section VIII: Rules for Construction of Pressure Vessels Division 1*

[7] American Petroleum Institute. *Management of Hazards Associated With Location of Process Plant Permanent Buildings*. Dec 2009. API RP 752.

[8] —. *Management of Hazards Associated with Location of Process Plant Portable Buildings.* Jan 2012. API RP 753.

[9] —. *Recommended Practice for the Design of Offshore Facilities Agains Fire and Blast Loading*. 2020. API RP 2FB.

[10] Simulia. *Abaqus / CAE User's Manual*. 2019.

[11] American Society of Civil Engineers. *Design of Blast Resistant Building in Petrochemical Facilities*. 2010. Vol 2nd Edition.

[12] Van den Berg, A.C. The multi-energy method: A framework for vapour cloud explosion blast prediction. *Journal of Hazardous Materials*, 1985; **12(1):** 1-10.

[13] TNO. *Application of correlations to quantify the source strength of vapour cloud explosions in realistic situations. Final report for the project: 'GAMES'*. October 1998. TNO Report PML 1998-C53.

[14] —. *Methods for the calculation of physical effects, (The Yellow Book) Part 1-2, 3rd Edition.* SDU Uitgevers, The Hague : s.n., 1997. CPR14E.

[15] Johnson D.M., Tam V.H.Y. Why DDT is the only way to explain some vapor cloud explosions. *Process Safety Progress*, 2017; **36 (3)**.

[16] Harris R.J., Wickens M.J. Understanding Vapour Cloud Explosions - An Experimental Study. Paper presented at: Institution of Gas Engineers 55th Autumn Meeting, Communication 1408; 1989; U.K.

[17] Davis, Scott, et.al. Large scale detonation testing: New findings in the prediction of DDTs at large scales. *Journal of Loss Prevention in the Process Industries*, 2017; **48(1)**.

[18] Steel Construction Institute. *Dispersion & Explosion Characteristics of Large Vapour Clouds*. 2015. Document ED024.

[19] Johnson, D. Michael, et al. Interpretation of damage caused by a vapor cloud explosion. *Process Safety Progress*, 2020; **39(1)**.

SPRING22
+18TH GCPS
A Joint AIChE and CCPS Meeting

# Potential Mitigations for LNG Pool Fires

**Jeffrey D. Marx**
**Quest Consultants Inc.**
**Norman, Oklahoma**
jdm@questconsult.com

**Benjamin R. Ishii**
**Quest Consultants Inc.**
bri@questconsult.com

Prepared for Presentation at
American Institute of Chemical Engineers
2022 Spring Meeting and 18th Global Congress on Process Safety
San Antonio, TX
April 10 – April 14, 2022

# Potential Mitigations for LNG Pool Fires

**Jeffrey D. Marx**
**Quest Consultants Inc.**
**Norman, Oklahoma**
**www.questconsult.com**
jdm@questconsult.com

**Benjamin R. Ishii**
**Quest Consultants Inc.**
bri@questconsult.com

**Keywords: LNG, Consequence Modeling**

## 0 Abstract

After its beginnings over 50 years ago in international import/export and peak-shaving applications, the use of liquefied natural gas (LNG) continues to expand into new applications. In recent years, many of these applications have involved the use of LNG as a vehicular fuel, creating the necessity of small and medium-sized facilities to serve these needs. While the LNG industry has for decades been concerned about the potential hazardous impacts of accidental releases, the limited real estate for smaller facilities has made siting of such facilities more difficult. Because LNG regulations require that the effects of accidental releases on offsite areas be controlled or minimized, LNG facilities often include an impoundment basin that limits the size of a vapor cloud or a pool fire. Even with a well-designed impoundment, a pool fire involving LNG is a potentially damaging event due to the burning properties of LNG. An impoundment fire is one of the more straightforward hazards to model – until mitigation measures are considered. In this analysis, several potential choices for pool fire mitigation are evaluated. The associated modeling methodologies are outlined, and the effects of the mitigations are presented. Considerations of cost, active versus passive mitigations, and potential reduction on thermal radiation impacts are discussed.

## 1 Introduction

The LNG industry in the U.S.A. stands out as the only hazardous material category for which consequence analysis calculations are required for siting the facility in relation to the public. Although this is not necessarily the same in other parts of the world, most countries require some sort of evaluation of the potential hazards associated with LNG, and often those are tied to facility siting through various regulations. Due to the inherent hazards of natural gas and LNG, as well as the refrigerants used in liquefaction processes, the primary hazards that must be considered are flammable vapor clouds and fires. While the flammable vapor cloud hazards dominate siting with respect to the property lines at most LNG facilities, fire hazards must still be evaluated. Additionally, LNG regulations often require the use of impoundments for potential liquid spills, including accidental releases of LNG. It is these impoundments that define many pool fires.

The sizing, placement, and design of impoundments intended to contain LNG is a unique engineering task. The impoundment must be designed to contain spilled LNG as well as rainwater, and simultaneously be sized and located to minimize (to the extent possible) vapor generation and the size of a pool fire should the vapors be ignited. These considerations, with the exception of controlling the thermal radiation from an LNG pool fire, will not be evaluated in this work; the focus is on mitigation methods for a fire that may occur at an existing facility.

## 2 Pool Fire Modeling

There are many choices for modeling of pool fires, from simple equation sets to computational fluid dynamics (CFD) approaches. Most simple calculation methods have been superseded by empirical solid-flame models due to their low computational costs and ease of use. These models represent the flame as a radiating solid by calculating the size, orientation, and radiation characteristics with testing-derived correlations and physical property data. This type of model can typically predict the radiation at a receptor location from burning pools of a wide range of flammable or combustible materials. Further considerations in this paper area based on such models.

One of the most well-known model for modeling pool fires is the LNGFIRE3 model[1]. This model is mandated by U.S. regulations and is widely used for facility siting purposes. While the model is easy to use, it is limited by the fact that it only considers land-based LNG pool fires and does not allow any modifications to account for mitigation measures. There are many other models available, in published literature or commercially available packages; these will not be discussed here. The model used in calculations presented here is the pool fire model contained in the CANARY by Quest® consequence modeling package. This model, like all models of its type, determines the radiative impact at locations around the flame by the following steps:

- *Define the geometric shape of the flame:* CANARY uses a tilted elliptical cylinder for circular pools or rectangular pools with aspect ratios up to 4:1.
- *Define the dimensions of the flame:* Correlations are applied to set the length and bending angle of the flame, based on the burning rate of the material, the size of the pool, the wind speed, and other variables. Flame drag in the downwind plane is also estimated.
- *Assign the surface emissive power (SEP) of the flame:* CANARY uses a two-zone model, with a clean-burning zone near the base of the flame and smoky zone above. Material properties, as well as the pool size, are also applied to determine the SEP.
- *Calculate the radiation from the flame to a receptor:* A piece-wise integration of the flame surface, combined with view factor determination and calculation of atmospheric transmittance, provides the thermal radiation flux at a potential receptor. Consideration of multiple receptor locations provides radiation isopleths around the flame.

This methodology (in CANARY and similar approaches in other models) has been applied to field-scale tests to validate and refine the models such that they provide good prediction of the potential thermal radiation impacts for pool fires in uncomplicated geometries.

# 3 Mitigation Options

Mitigation of pool fires is often valuable in the prevention of injuries to people, as well as avoidance of escalation due to the fire's effects. As the concern for both of these impacts is due to thermal radiation, it is important to understand how pool fires become a hazard. The emitted radiation is strongly dependent on the size of the flame, as defined by the fire base dimensions and the flame length or height. The flame height is primarily a function of the impoundment base dimensions and the rate of vapor production from the pool. Vapor production (which forms the vapors that burn in a flame) is dominated by heat input into the pool. For most materials, this heat input comes from radiation from the flame itself. For cryogenic materials, heat transfer from the substrate (impoundment) will also contribute to vapor generation. Thus, any measure that reduces the vaporization rate from the liquid pool will reduce the size of the flame and thus the thermal radiation received at a receptor.

Thermal radiation at a receptor is also dependent upon the distance between the flame and the receptor. The received thermal radiation flux decays with the square of the distance between flame and receptor. Additionally, anything that will block the line of sight from the flame to receptor can either eliminate or reduce the thermal radiation load to the receptor. This may include physical barriers or water spray systems.

## *3.1 Active vs. Passive Mitigation*

The first distinction for mitigation systems that needs to be made is the difference between active and passive mitigation. This is a much discussed topic in process safety[2] as well as fire protection disciplines. A mitigation classifies as passive if it requires no energy or signal to activate it. This is generally the preferred choice as compared to active mitigation, which requires an input or activation energy. Examples of each type are listed in Table 1, with further discussion in the subsequent sections for selected mitigation methods.

**Table 1. Potential Pool Fire Mitigation Methods**

| Active Mitigations | Passive Mitigations |
|---|---|
| Water | Distance/Spacing (Layout) |
| Dry Chemical | Impoundment |
| Medium/Low Expansion Foam | Foamglas® |
| High Expansion Foam | Physical Barriers |

## *3.2 Layout and Impoundment*

The use of distance between an impoundment and receptor is the first choice for passive mitigation. Receptors that are far enough away from a pool fire cannot be impacted. However, it is oftentimes not feasible to provide large distances between an impoundment and potential receptors. Impoundments are best placed near the potential spill points to avoid long transport trenches and the subsequent vaporization. Additionally, in small facilities, extra space is frequently not a luxury available to the facility designers.

As introduced above, impoundment design can be an important part of facility design, and so can contribute to mitigation in a passive manner. The primary concept that applies is the footprint or surface area of the impoundment. As the area decreases, the size of a fire in the impoundment decreases, thus reducing thermal radiation in the areas around the impoundment. While increasing the depth of an impoundment is not always feasible (for reasons not discussed here), it is always a potential mitigation factor in the design phases of a facility.

An additional factor that applies to cryogenic spills such as LNG is the material that the impoundment is constructed from. While the most common material is regular, high-density concrete, it is possible to reduce the heat transfer into the collected liquid with insulating concrete or specialized surface treatments. These are most effective in reducing conductive heat transfer driven vaporization and therefore the size of a flammable vapor cloud. Once a pool fire begins, such features are only minimally effective in reducing flame size.

### *3.3 Physical Barriers*

The placement of a barrier between a pool fire and a receptor, provided that the barrier is able to survive the radiative impact of the fire, provides a significant reduction in incident radiation. Barriers are passive mitigation, as it is impractical to actively move a barrier into place in response to a fire. If the barrier is large enough to eliminate the line of sight between receptor and the entire fire, the hazard is fully mitigated. If the barrier only partially obscures the flame, the level of mitigation is a function of view factors and can vary significantly with location relative to the flame. This behavior was addressed[3] with a pool fire modeling methodology to account for simple barriers between a pool fire and a receptor.

### *3.4 Water*

The classic active fire-fighting mitigation is water, but this is not always an effective mitigation. The modes of application of water for pool fire are numerous. Within the context of LNG pool fires, direct application of water to a pool fire may increase the hazard. Water cannot extinguish an LNG pool fire and the "heat" provided as water contacts LNG increases vaporization, making the fire larger.

Equipment or vessels in the area of an LNG impoundment can be protected by a deluge or water spray system designed for that specific piece of equipment, as long as the water does not drain to the same impoundment where LNG is collected. The application of water provides a heat sink for any incident thermal radiation, protecting the equipment or vessel.

As a barrier mitigation, water curtains (or mist systems) are also an option. In this application, one or more spray nozzles are set up to create a wall of water droplets between the fire and the receptor. Water absorbs a significant amount of the emitted thermal radiation, thus providing the desired protection, but does not provide the level of protection that a solid barrier would. The primary nuance to water curtains is that their effectiveness is based on how completely they can block the line of sight between the receptor and all parts of the flame. This can be difficult to achieve for large pool fires, especially when considering variations in wind.

### *3.5 Dry Chemical*

In a series of tests conducted in the late 1970's and 1980[4], the efficacies of various dry chemicals and high expansion foams were tested on LNG, LPG, and other pool fires. Dry chemical, when properly applied, can extinguish any type of pool fire. However, as the pool size grows, the application rate, application technique, impoundment depth, and attack points become more important. For fires larger than about 6 x 6 meters (20 by 20 feet), consistent coverage becomes much more important, as unextinguished pockets of flame tend to re-ignite the pool fire. This is exacerbated by the presence of any obstacles in the pool.

In application to LNG impoundment fires, dry chemical is a viable active mitigation system. Fires can be extinguished by a fixed application system, or by trained personnel (fire fighters), provided that sufficient dry chemical is available when needed. One concern with fire extinguishment is that the remaining LNG (or other material) would then continue to vaporize, creating a flammable vapor cloud that, if re-ignited, could create a flash fire (introducing a new hazard) and would certainly re-establish the pool fire.

### *3.6 Foam Systems*

A decades-old pool fire mitigation is the addition of a foaming agent to water. Foams are available in low-, medium-, and high-expansion varieties. The differentiation between these varieties is the ratio of the foam volume to the original water solution volume. While effective for other fire types and non-LNG pool fires, low- and medium-expansion foams have generally been found to be ineffective for LNG fires. This is partially due to the fact that the foam is mostly water. When the foam breaks down it thus adds water to the LNG pool, making the fire larger.

High-expansion foam is the most effective for LNG fires, achieving an expansion ratio of approximately 500:1. Foam is applied by spraying a solution of water and foaming agent onto a screen over which air is forced. This creates many small bubbles, which forms a thick blanket that will sit on top of the LNG pool. Foam cannot ever fully extinguish an LNG fire, so the foam breaks down over time and must be re-applied. Foam application also requires that the foam generators be relatively close to the LNG pool.

High expansion foam offers a means to control, but not extinguish, an LNG fire. As a mitigation choice, it provides the desired fire control (reduction in radiant heat flux around the pool) but must be continuously re-applied to remain effective. Additionally, foam systems come with the cost of the foam solution, a physical system to deliver water and solution, means to apply the foam to the correct location, and requirements for periodic testing of the system.

### *3.7 FoamGlas®*

One of the more effective passive mitigation methods is a product marketed by Owens Corning: Foamglas® PFS (Pool Fire Suppressant)[5], a closed cell rigid insulation manufactured from glass. Foamglas is inert to most chemicals, remains rigid to temperatures of 1,346 °F (730 °C), is insect and vermin resistant, and suitable for most weather conditions. Foamglas is manufactured with a bulk density of 120 kg/m$^3$ (7.5 lb/ft$^3$), allowing it to float on a pool of liquefied natural gas (LNG), which typically has a density of about 450 kg/m$^3$ (28 lb/ft$^3$). Foamglas is installed in the impoundment where LNG is expected to be collected. In the event of an LNG spill, the Foamglas remains on the top of the LNG, and if the vapors are ignited, blocks the flame radiation that would normally reach the LNG and thus controls the size of the fire.

The original Foamglas was provided as a set of cubes in polyethylene bags for environmental protection. The bags are to be placed in the impoundment two or three layers thick to provide the best mitigation. The Foamglas® PFS System (Gen 2) features stainless steel shrouding over the Foamglas cubes to protect them from environmental factors and aid in keeping them aligned in the impoundment basin. This also helps to form a physical barrier between the flames and the LNG, while still floating on the pool, further aiding reduction of the vaporization rate.

# 4 Modeling Fire Mitigations

When modeling pool fires, the first-pass approach is to ignore mitigation measures, especially active mitigations, under the assumption that they may not function as desired. While this produces a conservative result, it is still valuable to understand the effects of not having any mitigation measures for a pool fire.

Many models do not offer any explicit provision for mitigation measures, and some (e.g., LNGFIRE3) offer no ability to change the related modeling parameters. Thus, it is incumbent on the user to understand the workings of the model and how to implement the modeling to accommodate the effects of any mitigations. This first requires an understanding of how a given mitigation measure would affect a pool fire.

## *4.1 Research on Suppression of LNG Fires*

There have been several research programs to investigate pool fires[6], and a few that investigated the efficacy of pool fire mitigations. The notable tests involving high expansion foams and Foamglas blocks, on which pool fire calculations can be based, are described below.

### 4.1.1 High Expansion Foam

In the tests conducted by Johnson et.al.[4], the efficacy of various high expansion foams was tested on LNG, LPG, and other pool fires. It was demonstrated that control of the fire could be achieved with sufficient application of foam, including re-application as the foam collapsed or was moved by the wind. Control was defined as reduction of the measured thermal radiation of approximately 75%. In a similar test conducted by Shell[7], LNG was introduced into a 6 x 6 meter pool and after a delay, ignited. The pool fire was allowed to burn so that the flame was established, and subsequently, a 500:1 high expansion foam was applied. Radiometer measurements showed a reduction of measured thermal radiation flux up to about 95%.

More recently, the use of high-expansion foam to suppress LNG fires[8] confirmed the behavior of a 500:1 foam on the suppression of LNG pool fire radiation: measured flux levels dropped to below 10% of the un-suppressed fire within a few tens of seconds after application began. While the effects were shown to be dependent on foam application rate and were dependent on maintaining a foam blanket on the LNG, foam was shown to be an effective mitigation.

4.1.2 Foamglas

A study conducted at the Shell Thornton Research Centre in 1981[9] in a 6 x 6 m impoundment using a 90-100 mm (3.5 to 3.9 inch) layer of Foamglas blocks placed in the impoundment floor resulted in a peak thermal radiation intensity of 0.5 $kW/m^2$ (159 $Btu/hr\text{-}ft^2$) compared to an unmitigated peak intensity of 9 $kW/m^2$ (2,855 $Btu/hr\text{-}ft^2$). This is a reduction in peak radiation intensity of slightly over 95%. Tests were also conducted in a 1.8 m diameter tray with a 200 mm (7.9 inch) layer of small cellular glass cubes. The resulting reduction in thermal radiation was approximately 97%.

In Suardin's PhD dissertation[8], the use of Foamglas to suppress LNG vapor production and fire thermal radiation from LNG in an impoundment was evaluated in addition to high expansion foam. The study concluded that Foamglas was effective in suppressing LNG pool fires and the reductions replicated the 1981 study well.

A more recent study[10] presented the results of an LNG vapor and fire control study in a 2 x 2 m pit using the 2$^{nd}$ Generation Foamglas. The reported radiant heat flux reduction for this test was 90%.

### *4.2 Matching Test Data with Pool Fire Simulations*

In order to be able to model an LNG pool fire when either high expansion foam or Foamglas blocks are used, the first task is to match the available data presented from the tests. Depending on the model being used, the methodology may take different forms. For the LNGFIRE3 model, this cannot be done. For other models, variables such as flame height can be forced in the model to simulate mitigation measures. In the CANARY model, this takes the following form:

1. Model the test's scenario to validate that the model matches the unmitigated scenario.
2. Modify the internally-applied pool fire burning rate (not a user-defined parameter) until the radiation received at the receptor location is reduced by the amount measured in the tests (e.g., 90% of the measured value).
3. Apply the resulting burning rate to the proposed impoundment scenario.

When this methodology is applied to the various tests described above, estimates for the appropriate modifications to a pool fire model can be made. Although the variable in the CANARY model is the burning rate, similar modifications in other models that affect the resulting flame height can be determined with a similar methodology. It is worth noting that any approach that does not result in a significantly shorter flame will not properly represent the mitigated pool fire when high expansion foam or Foamglas blocks are considered for mitigation measures.

Table 2 presents the results of the test matching by modification of CANARY's burning rate.

**Table 2. Mitigation Modification Results**

| Test | Mitigation Type | Unmitigated CANARY Predicted Flame Length [m] | Modified Burning Rate [kg/m²-s]† | Mitigated CANARY Predicted Flame Length [m] |
|---|---|---|---|---|
| Shell (Lev)[7], Test #3 | High Expansion Foam | 16.5 | 1.08E-03 | 1.0 |
| Suardin[8], Test #4A | High Expansion Foam | 20.5 | 8.65E-04 | 1.1 |
| Shell (Lev)[9] | Foamglas® | 16.4 | 2.16E-03 | 1.5 |
| RPI[10] | Foamglas® Gen2 | 7.8 | 4.34E-03 | 1.1 |

† For reference, CANARY's standard burning rate for an LNG mixture is 0.108 kg/m²-s

### *4.3 Modeling Challenges*

As seen in table 2, there is not a clear value to be assigned to a modified burning rate when applying this methodology. Factors applied to the CANARY burning rate range from 1/25th to 1/125th. There are several reasons for these findings:

- The test data is (naturally) highly variable, even within one test. Radiant heat flux varies with time due to flame fluctuations, wind speed and direction changes, and burning rate changes in the pool of LNG.
- Comparison for mitigation options always involves two tests, or two phases of one test. Differences are necessarily introduced with this process.
- The thermal radiation flux reductions from high expansion foam are also variable over time, as the foam application and its breakdown create a continuously dynamic situation.
- Most test series feature a small set (sometimes one) of fire scenarios. Furthermore, the details of every test are not always presented in literature. Thus, the sample set for evaluation is small.

In addition to the difficulties in estimating the thermal radiation reduction due to mitigation using published test data, there are uncertainties in application of these derived values to real problems. Chief among these is the size of the LNG pool. The available tests with mitigation measures represent impoundments between 2 meters square and 6.5 x 10 meters. While some impoundments (especially at smaller facilities) may match these sizes, the impoundments at many LNG facilities are necessarily larger. It is not clear what the effects of a larger pool may be, especially with the application of high expansion foam.

Finally, there are uncertainties with pool fire modeling at very low burning rates. While the performance seems to match the mitigated test data reasonably well, there is not a comprehensive set of radiometer data on which to validate large pools with very short flames. Thus, the model construction may not be well suited for such flames, creating uncertainty in the use of modification factors as developed in Section 4.2.

The evaluation presented above seems to show that high expansion foam would provide a greater thermal radiation mitigation than Foamglas. While this is true for a short time period, the variability over time, including the need to constantly re-apply foam, does not support this as a definitive advantage. The passive nature of Foamglas, and its ability to remain in place for the duration of the fire, provide advantages over high expansion foam.

Due to the above factors, for the methodology presented above (which is specific to the CANARY model), a burning rate modification (multiplier) of 0.05 ($1/20^{th}$) is recommended for either mitigation method. This choice, while slightly more conservative than the results obtained in the experimental comparisons, attempts to account for the challenges of modeling either mitigation, the variability in time for high expansion foam, and the uncertainties of larger pool fires than were involved in the referenced experiments.

## 5 Example Problem

To demonstrate the effects of pool fire mitigation, consider an LNG facility with an LNG spill impoundment measuring 10 x 12 meters and sufficiently deep to contain a spill from the largest system it serves. To compare the mitigated vs. non-mitigated scenario, a thermal radiative flux of 12.6 kW/m$^2$ will be used as the modeling endpoint. This value is often linked to escalation during prolonged exposure –lasting for tens of minutes. Escalation could include damage to cabling, instrumentation, structural steel (beginnings of loss of strength), or heating of process equipment such that they relieve their contents.

Thermal radiation isopleths for this example scenario, in both the unmitigated and mitigated fire scenarios, are demonstrated in Figure 1, where the fire in the 10 x 12 meter impoundment is modeled with a 5 m/s wind blowing along its shorter dimension. The mitigated flame (using either high expansion foam or Foamglas) is predicted to have a length (height) of 4 meters as opposed to 25 meters in the unmitigated case.

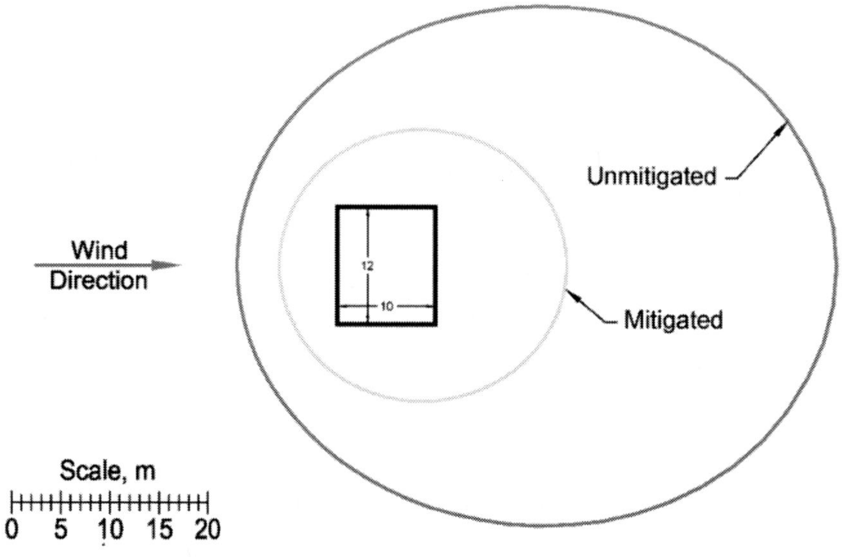


**Figure 1. Thermal Radiation Isopleths for 12.6 kW/m$^2$ Impacts from an LNG Pool Fire During 5 m/s Winds**

## 6 Concluding Remarks

The use of mitigation for potential LNG pool fires is an important part of fire protection and facility design. Of the mitigation choices that apply to the pool fire itself, many cannot be easily evaluated with pool fire modeling, or may require a detailed 3-dimensional approach. Two mitigation methods were evaluated: high expansion foam and Foamglas blocks. While the implications of implementation for these choices have their own unique benefits and drawbacks, both choices offer significant reduction in the thermal radiation flux that a receptor may experience near an LNG pool fire. The modeling methodologies explored in this paper are relatively straightforward and can be applied to most pool fire impoundment scenarios to evaluate the thermal radiation reductions that could be realized with high expansion foam or Foamglas block applications.

## 7 References

[1] GRI, *LNGFIRE: A Thermal Radiation Model for LNG Fires*. Gas Research Institute, GRI-89/0176, June29, 1990.

[2] Spitzenberger, et.al., Strike the Right Balance between Active and Passive Fire Protection, *Chemical Engineering Progress*, American Institute of Chemical Engineers, p. 61-68, June 2016.

[3] Gavelli, F., The Effect of Barriers on Reducing Thermal Heat Fluxes from a Hydrocarbon Pool Fire. *Journal of Loss Prevention in the Process Industries*, Vol. 72, 2021, 104554.

[4] Johnson, et.al., Control and Extinguishment of LPG Fires. U.S. Department of Energy, Report No. DOE-EV-06020-T3, August 1980.

[5] Foamglas PFS. https://www.foamglas.com/en-us/advice-center/general-advice/cryogenic-spillage-protection Accessed January 2022.

[6] Suardin, J.A, et.al., Field Experiments on High Expansion (HEX) Foam Application for Controlling LNG Pool Fire. *Journal of Hazardous Materials*, Volume 165, p.612-622, 2009.

[7] Lev, Y., Some Experiments on the Use of Foam for LNG Fire Fighting, *Fire Technology*, Volume 17, p. 17-25, February 1981.

[8] Suardin, J.A. The Application of Expansion Foam on Liquefied Natural Gas (LNG) to Suppress LNG Vapor and LNG Pool Fire Thermal Radiation. A dissertation submitted to the Office of Graduate Studies of Texas A&M University, August 2008.

[9] Lev, Y. Novel Method for Controlling LNG Pool Fires, *Fire Technology*, Volume 17, p.275-284, November 1981.

[10] RPI. *Vapour & Fire Control Testing of FOAMGLAS® PFS System (Gen 2) on LNG*, Final Report RPI Ref P1262, Resource Protection International, February 2014.

# 640075 Study on Mass Burning Rate of LNG Pool Fire Using a Validated CFD Model

**Abstract:**

Liquefied natural gas (LNG) is widely used, because it provides an easy and economic solution to transport and store natural gas, especially when pipelines are not available. Pool fire is the major hazard of the LNG accident according to the report of the U.S. Government Accountability Office, because LNG fire has a large surface emissive power compared with other hydrocarbon fuels, which may lead to domino accident. Previous works have developed Computational Fluid Dynamics (CFD) models of LNG pool fire, but the mass burning rate was fixed manually as the model input, ignoring the fact that the mass burning rate is determined by the fuel's physical properties and heat input to vaporize the fuel. In this work, a fire dynamics simulator (FDS) model of LNG pool fire controlled by material physical properties was developed and validated against a series of LNG pool fire experiments carried out by Mary Kay O'Connor Process Safety Center (MKOPSC). Statistical performance measurement shows that the model is superior to the semi-empirical formula in predicting the mass burning rate of LNG pool fire under different pool sizes. A flame geometry analysis software was developed by comparing the different algorithms, and the centroid method was selected. The results show that the fire model (200 $kW/m^3$ as flame contour) can accurately predict the flame geometry in the experiment. The influence of wind speed and dike height on the mass burning rate was also studied by this model. The results show that the height of concrete dike is negatively correlated with the mass burning rate of LNG pool fire. The effect of wind speed on LNG mass burning rate is complicated. The forced convective boundary layer at lower wind speed will promote LNG combustion and increase the mass burning rate, while higher wind speed will reduce the mass burning rate by reducing thermal radiation feedback.

*Corresponding author: Bin Zhang, bzhang@njtech.edu.cn;

SPRING22
+18TH GCPS
A Joint AIChE and CCPS Meeting

# Risk Based Approach for Determining Fireproofing Requirements

**Raminaidu Girada**
**Senior Process Safety Engineer, Dragon Oil**
**Executive Building 16, 5th Floor, Health Care City, Dubai**
**rgirada@dragonoil.com**

**Nino Nicotra**
**Principal Process Safety Specialist, Bechtel Energy**
**3000 Post Oak Boulevard, Houston, TX 77056**
**anicotra@bechtel.com**

Prepared for Presentation at
American Institute of Chemical Engineers
2022 Spring Meeting and 18th Global Congress on Process Safety
San Antonio, TX
April 10 – 14, 2022

# Risk Based Approach for Determining Fireproofing Requirements

**Raminaidu Girada**
**Senior Process Safety Engineer, Dragon Oil**
**Executive Building 16, 5th Floor, Health Care City, Dubai**
**rgirada@dragonoil.com**

**Nino Nicotra**
**Principal Process Safety Specialist, Bechtel Energy**
**3000 Post Oak Boulevard, Houston, TX 77056**
**anicotra@bechtel.com**

**Keywords:** Liquefied Natural Gas (LNG), LNG, Pool Fire, Risk Based, Cost Savings

## Abstract

There is only one known recommended practice, namely API RP 2218, that provides guidelines for selection and application of fireproofing in the petroleum and petrochemical processing plants. This recommended practice specifically addresses "pool fire" scenarios. The fireproofing need is determined based on equipment that handles flammable liquid above a specific threshold volume and the extent of the fire-scenario envelope around the fire potential equipment is suggested by API RP 2218. The same document also suggests performing a risk assessment in determining the potential needs of fireproofing. The risk-based approach is particularly useful when dealing with LNG and liquefied gases used as refrigerants, because their accidental release seldom forms a liquid pool due to their very low boiling point and high evaporation rate. A recent study on pool fire modelling using an industry approved modelling tool proved that most of the liquid holding equipment in an LNG plant did not form liquid pools after an accidental release. Some other scenarios resulted in pool fires that did not last longer than a few minutes and did not cause harm to structural steel members. The steel members can withstand fires for several minutes based on their dimensions. This study shows how calculated pool fire size and duration will play a significant role in determining fireproofing requirements, and how a risk based fireproofing design together with the calculated time to failure of structural steel members may result in a significant reduction of fireproofing material quantities, and consequently a cost saving, which is desirable in the current economy.

# 1 Introduction

Currently in LNG plants, fireproofing is considered for equipment supports, pipe rack support and other auxiliary critical hydrocarbon carrying pipe supports within a fire proofing zone to improve its structural integrity during a fire. The principal value of fire proofing is realized during the early stages of a fire by providing additional time to fire fighters to control or extinguish a fire thus by delaying additional loss of containment which may escalate the incident. The fireproofing need is determined based on equipment that handles flammable liquid above a specific threshold volume and the extent of the fire-scenario envelope around the fire potential equipment is suggested by API RP 2218. Per the API, the fireproofing zone is defined for every fire potential equipment, that extends 20 ft. to 40 ft. (6 m to 12 m) horizontally and 20 ft. to 40 ft. (6 m to 12 m) vertically from the source of liquid fuel. As a typical approach most projects consider fireproofing zones that extend 30 ft. (9m) horizontally and 30 ft. (9m) vertically from the source.

The above qualitative approach does not reflect the realistic possibility of pool fire scenario for releases from equipment handling liquefied gases (i.e. refrigerants) and LNG. Therefore, a consequence analysis may be performed to model when a pool fire is formed, what is the diameter, the flame length, and for how long it burns. Moreover, fire endurance/resistance times of unprotected steel structural members shall also be estimated since short duration fires will not harm structural steel members. This analysis has been performed on an LNG Project to determine the fireproofing needs for structural supports of piping and equipment handling refrigerants and LNG.

# 2 General Technical Guidance

## *2.1 Pool Fire Formations*

A pool fire may result via a number of scenarios, and it begins typically with the release of flammable material from process equipment. If the material is liquid, stored at a temperature below its normal boiling point, the liquid will collect in a pool. The geometry of the pool is either dictated by the surroundings (i.e., diking), or unconstrained in an open flat area, which can happen when the liquid quantity spilled is inadequate to completely fill the diked area or is unconfined. If the liquid is stored under pressure above its normal boiling point, then a fraction of the liquid will flash into vapor, with un-flashed liquid remaining to form a pool in the vicinity of the release.

The pool of flammable liquid may be ignited creating a pool fire. If the release of flammable material from a pressurized process equipment continues, then a jet fire is also likely. If the ignition source occurs at the very beginning of the release, then inadequate time is available for the liquid to form a pool and only a jet fire will result.

Pressurized liquefied gases such as LNG, methane, ethylene and propane will in general disperse as mists or vapor, but may possibly form a small pool if the source operates at low pressure conditions. An exception has to be made for butanes, pentanes and heavier hydrocarbons, which always create pool.

Refer to the following **Figure 1** for pictorial view of the pool fire formation.

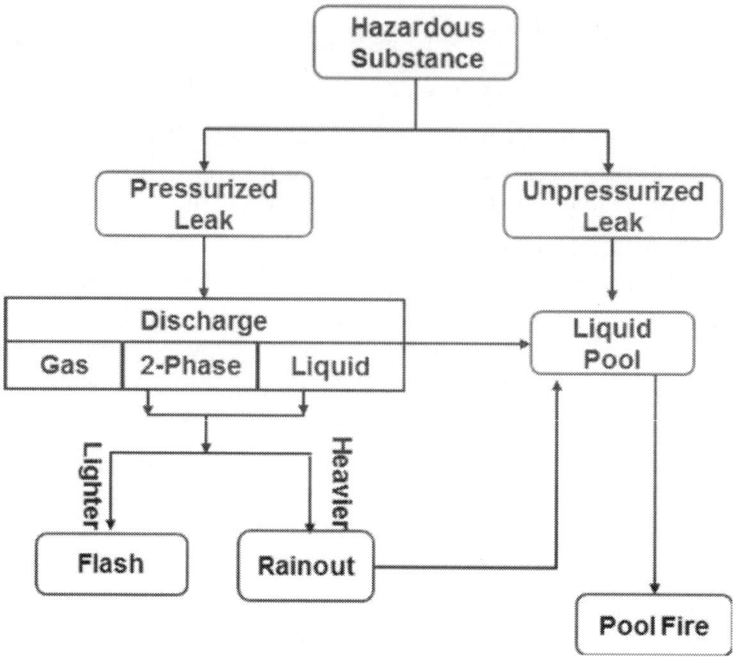


**Figure 1: Pool Fire Formation**

### *2.2 Determination of Pool Fire scenario*

This section provides an overview of the methodology for determining pool fire existence.

Accidental release of flammable materials such as Pentanes and heavier hydrocarbons create liquid pool irrespective of their operating conditions. However, the lighter hydrocarbons such as butane and propane may create liquid pool only under specific operating conditions. The potential for creating liquid pool is based on certain parameters as described below:

- **Butanes**

  Butane releases potentially create liquid pool except for high pressure releases and slight sub cooling of release. Further consequence analysis is required to check if a liquid pool is created from the release.

- **Propane and Ethylene**

  Ethylene and propane liquids may or may not pool, depending on conditions. It has been found that a pool formation can be correlated on several parameters: (1) vessel operating pressure, (2) hole size of leakage, (3) the amount of subcooling, which is also important, but the assumption is the liquid will be slightly subcooled, and (4) confinement of *fire hazard area*. Together, these variables establish the size of the pool.

  Higher pressure releases are more likely to vaporize and mist, and low pressure releases are more likely to pool. Based on a slightly subcooled liquid at pressure, the analysis is divided into three regimes (1) no liquid pools, (2) highly likely to form a significant pool fire, and (3) possible pool fire. Depending on the size of the release, slope, temperature and composition, a consequence model may be necessary to determine if a pool fire is warranted.

- **LNG**

  Similar to propane and ethylene liquids, LNG may or may not pool depending on several parameters: (1) vessel operating pressure, (2) size of rupture hole, (3) the amount of subcooling, and (4) confinement.

Therefore, a quantitative approach was performed in the following sections.

# 3 Leak/Release Frequency

Releases from small hole sizes up to 2" (50mm) are more likely to occur than larger hole sizes. The HSE in UK has published Hydrocarbon Release Database (HCRD) with over 20 years of incident data, including releases of hydrocarbon, as well as the data for a number of years equipment in service.

Additionally, a paper [2] published in 2016, used HCRD to evaluate the overall frequency distribution for the two-inch hole size (50 mm hole size). For equipment (piping, compressors, pumps, heat exchangers, valves, filters, pressure vessels, storage tanks, and pig launchers/receivers), hole sizes up to (50 mm) represent 95.46% of the total failure rate.

## *3.1 Case Reference*

For a specific project, a QRA was performed, and release frequencies were calculated using the historical data from the UK HSE Offshore Hydrocarbon Release Database. The following tables presents the hole size and leakage distribution.

| Representative Hole Size | Hole Size Range |
|---|---|
| 2mm | 0 – 3.7mm |
| 7mm | 3.7 –12.4mm |
| 22mm | 12.4 – 39.2mm |
| 70mm | 39.2 – 102.4mm |
| 150m | >102.4mm |

| Representative Hole Size | Leaks / Year | Percentage | Cumulative |
|---|---|---|---|
| 2mm | 3.15 | 76.6% | 76.6% |
| 7mm | 0.62 | 15.1% | 91.7% |
| 22mm | 0.28 | 6.7% | 98.4% |
| 70mm | 0.04 | 1.0% | 99.5% |
| 150mm | 0.02 | 0.5% | 100.0% |
| *Total Leaks* | *4.10* | *100%* | |

These tables show how most releases happen for small hose sizes up to 70mm with a cumulative frequency of 99.5%, where the category 70mm groups all hole sizes ranging between 39.2mm and 102.4mm.

Therefore, 50 mm hole size at various isolatable sections can be acceptable as the basis for the pool fire modeling.

# 4 Pool Fire Modeling

For pool fire modeling of the LNG Project, the following conditions would apply:

- A release from flammable equipment has been modeled using an industry recognized software tool. Flammable fluids modelled: methane, propane, ethylene, and LNG
- Pool fire resulting from 50mm equivalent diameter ruptures have been considered as the basis for fireproofing since these are more likely than bigger or full-bore ruptures [1][2] as described in the above section.
- The pool fire have been modelled for different weather conditions (F 1.5/s, D 3m/s, D 5m/s). Since the stability class D and wind speed 5m/s provided more severe results, these have been considered as a basis for the fireproofing.
- Late Pool Fires (LPF) resulting from a delayed ignition have been considered for determining the fireproofing since the results obtained are more severe than those for Early Pool Fires.

## *4.1 Pool Fire Results*

The following table is a snapshot of the results which have been obtained from the Pool Fire Modeling:

| Area | Isolatable Section/ Equipment | Release Rate [kg/s] | Leak Duration [s] | Total Inventory Released [kg] | Weather | Pool Diameter [m] | Flame Length [m] | Pool Fire Duration [s] |
|---|---|---|---|---|---|---|---|---|
| 202J | 2V-1413 | 3.8 | 57 | 218 | D 5m/s | 0.4 | 1.1 | 65142.3 |
| 202H | 2E-1504 | 32.0 | 3397 | 108622 | D 5m/s | 7.2 | 14.9 | 38074.4 |
| 202H | 2E-1503 | 51.8 | 347 | 17988 | D 5m/s | 4.3 | 10.5 | 17281.4 |
| 202J | 2V-1412 | 1.9 | 58 | 111 | D 5m/s | 0.6 | 1.6 | 12685.7 |
| 202G | 2V-1602 | 42.2 | 361 | 15216 | D 5m/s | 5.0 | 11.5 | 11038.9 |
| 202J | 2V-1404 | 45.5 | 1199 | 54576 | D 5m/s | 8.9 | 23.7 | 7442.4 |
| 208C | 2V-2002A | 37.0 | 3600 | 133363 | D 5m/s | 18.9 | 40.3 | 3969.5 |
| 208C | 2V-2001A | 22.0 | 2624 | 57628 | D 5m/s | 18.9 | 29.1 | 2920.6 |
| 202F | 2P-1703 | 34.2 | 101 | 3467 | D 5m/s | 4.2 | 14.1 | 2095.3 |
| 202J | 2E-1404A | 57.1 | 1496 | 85359 | D 5m/s | 20.9 | 43.2 | 2077.7 |
| 208C | 2E-2001 | 1.4 | 20 | 28 | D 5m/s | 0.6 | 2.5 | 1647.1 |
| 202F | 2V-1706 | 22.2 | 322 | 7140 | D 5m/s | 7.2 | 20.5 | 1471.3 |
| 202G | 2V-1604 | 10.4 | 1720 | 17967 | D 5m/s | 11.8 | 28.1 | 1437.0 |
| 202J | 2E-1403 | 55.7 | 1036 | 57666 | D 5m/s | 21.7 | 44.4 | 1295.6 |
| 202G | 2V-1603 | 23.8 | 540 | 12865 | D 5m/s | 11.3 | 26.9 | 1159.0 |
| 208C | 2V-2001A | 22.0 | 2624 | 57628 | D 5m/s | 32.8 | 42.7 | 967.8 |
| 202J | 2E-1402 | 54.6 | 687 | 37555 | D 5m/s | 22.0 | 44.7 | 826.4 |
| 202J | 2V-1502 | 51.2 | 510 | 26142 | D 5m/s | 23.9 | 34.3 | 823.9 |
| 202G | 2PK-1602 | 39.9 | 86 | 3434 | D 5m/s | 8.2 | 19.5 | 689.9 |
| 207C | 2V-2401 | 2.4 | 872 | 2116 | D 5m/s | 6.8 | 16.1 | 687.2 |
| 202G | 2E-1608 | 10.6 | 199 | 2116 | D 5m/s | 6.9 | 16.3 | 663.6 |

| Area | Isolatable Section/ Equipment | Release Rate [kg/s] | Leak Duration [s] | Total Inventory Released [kg] | Weather | Pool Diameter [m] | Flame Length [m] | Pool Fire Duration [s] |
|---|---|---|---|---|---|---|---|---|
| 208C | 2V-2002A | 37.0 | 3600 | 133363 | D 5m/s | 46.2 | 75.0 | 663.1 |
| 202F | 2E-1704 | 37.1 | 89 | 3309 | D 5m/s | 7.7 | 21.3 | 605.4 |
| 202G | 2P-1602A | 14.0 | 149 | 2093 | D 5m/s | 7.3 | 17.4 | 558.6 |
| 202H | 2P-1501 | 22.0 | 121 | 2650 | D 5m/s | 9.6 | 18.1 | 525.3 |
| 202J | 2E-1506 | 51.2 | 43 | 2199 | D 5m/s | 9.4 | 17.9 | 451.7 |
| 202F | 2E-1708 | 36.8 | 82 | 3029 | D 5m/s | 18.9 | 19.0 | 308.7 |
| 202F | 2V-1711 | 105.3 | 369 | 38861 | D 5m/s | 41.9 | 69.3 | 238.9 |
| 208C | 2V-2301 | 19.4 | 3600.0 | 69746 | D 5m/s | 59.0 | 88.0 | 216.1 |
| 202F | 2V-1710 | 90.4 | 392 | 35434 | D 5m/s | 47.0 | 75.1 | 173.5 |
| 202G | 2PK-1601 | 10.4 | 393 | 4110 | D 5m/s | 15.6 | 36.1 | 171.9 |
| 202G | 2E-1609 | 15.1 | 143 | 2159 | D 5m/s | 12.9 | 30.4 | 142.2 |
| 202H | 2PK-1501 | 51.8 | 91 | 4712 | D 5m/s | 25.6 | 35.9 | 130.3 |

**Table 1: Snapshot of the Pool Fire Results**

The above results show that the pool diameter and the pool fire duration are different for different equipment. The pool diameter is in the range of 0.1m to 35m, and the pool fire duration is in the range of 30 seconds to 124 minutes. In some case, the pool diameter is more than 10m, however, the pool fire duration is less than two minutes.

# 5 Fire Endurance Time of Unprotected Steel Columns and Beams

Fire endurance/resistance times have been calculated for unprotected columns and beams based on empirical equations provided in Society of Fire Protection Engineers (SFPE), Handbook for Fire Protection Engineering, Volume 2, 5th Edition, Table 53. 3.

## *5.1 Fire Endurance/Resistance Time Calculation Equations*

The empirical equations to calculate the fire resistance/endurance time is reproduced below from the SFPE Handbook [3] for Fire Protection Engineering.

$$\mathbf{R = 10.39\,(W/D)^{0.7}, \text{ for } W/D<10} \quad (1)$$

$$\mathbf{R= 8.3\,(W/D)^{0.8}, \text{ for } W/D\geq 10} \quad (2)$$

**(for critical temperatures of 1000°F)**

R = fire endurance time (min)
W = weight of steel section per liner foot (lb/ft), W=490 lb/ft$^3$[3]
D = heated perimeter (in.)

Heated Perimeter for Columns is calculated as below:

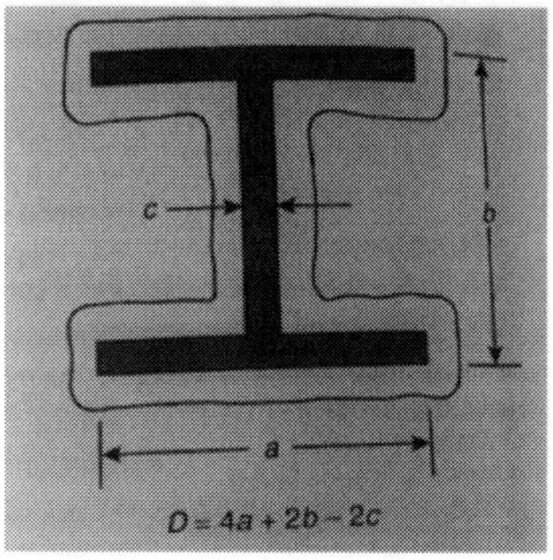


**Heated Perimeter for Columns, D = 4a+2b-2c** .......... (3)

Heated Perimeter for Beams is calculated as below:

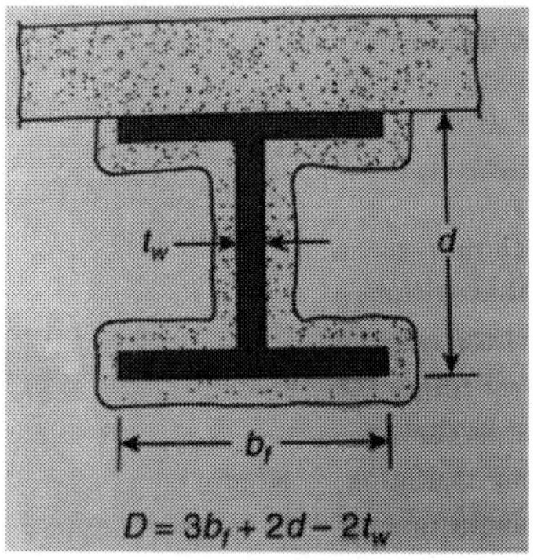


**Heated Perimeter for Beams, D = 4a+2b-2c** ............. (4)

In Equation (1), **weight of steel section per liner foot, W**, is calculated as below

**W = Area of Shape (A) x Density**................................................. (5)
Density of Steel = 490 lb/ft$^3$ [3]

In Equation (5), **Area of Shape (A)**, is calculated as below:

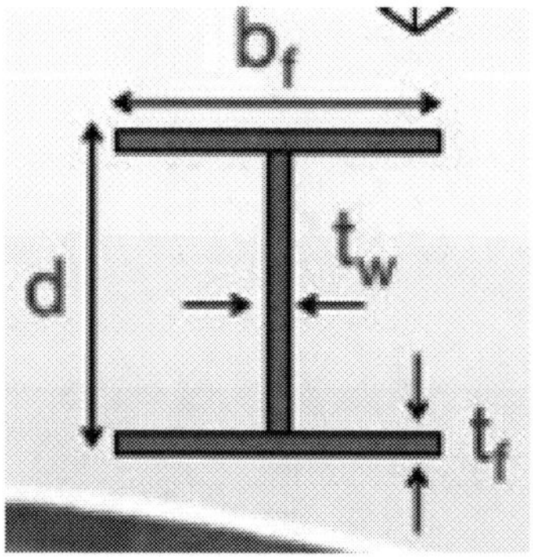


**$A=2(b_f)(t_f)+(d-2f)t_w$** ..........................................(6)

## *5.2 Fire Endurance Times*

The calculated Fire Endurance Times are tabulated below for steel columns and beams which are present in the areas where pool fire potential equipment is existing. In an area, if all the columns are of similar size, a single steel column has been considered as representative to calculate the endurance time. If there are different size columns, a single column from each size group has been considered as representative. The same method has been applied to beams as well to calculate the endurance times.

| Area | Column Dimensions | Density of steel (lb/ft3) | Area of Shape (ft2) | W (lb/ft) | D (in.) | W/D | Fire Endurance Time (min) |
|---|---|---|---|---|---|---|---|
| 202J | HW428X407X20X35 | 490 | 0.384 | 188.03 | 96.22 | 1.95 | 16 |
| | HW400X400X13X21 | 490 | 0.231 | 113.387 | 93.465 | 1.213 | 12 |
| | | | | | | | |
| 202A | WH900X400X16X36 | 490 | 0.453 | 222.23 | 132.598 | 1.676 | 15 |
| | | | | | | | |
| 202C | HW300X300X10X15 | 490 | 0.126 | 61.835 | 70.079 | 0.882 | 9 |
| | HW250X250X9X14 | 490 | 0.097 | 47.555 | 58.347 | 0.815 | 9 |
| | | | | | | | |
| 202F | HW428X407X20X35 | 490 | 0.384 | 188.413 | 96.221 | 1.958 | 16 |
| | WH1100X400X16X40 | 490 | 0.520 | 255.375 | 148.347 | 1.721 | 15 |
| | HW350X350X12X19 | 490 | 0.183 | 90.079 | 81.732 | 1.102 | 11 |
| | HN400X150X8X13 | 490 | 0.074 | 36.425 | 54.488 | 0.669 | 8 |
| | | | | | | | |
| 202G | WH900X400X14X40 | 490 | 0.468 | 229.795 | 132.756 | 1.731 | 15 |
| | HW300X300X10X15 | 490 | 0.126 | 61.835 | 70.079 | 0.882 | 9 |
| | | | | | | | |
| 202H | HW300X300X10X15 | 490 | 0.125 | 61.835 | 70.079 | 0.882 | 9 |
| | | | | | | | |
| 202K | HW400X400X13X21 | 490 | 0.231 | 113.38 | 93.465 | 1.213 | 12 |
| | HW300X300X10X15 | 490 | 0.126 | 61.835 | 70.079 | 0.882 | 9 |
| | | | | | | | |
| 202L | HW428X407X20X35 | 490 | 0.384 | 188.41 | 96.221 | 1.958 | 16 |
| | | | | | | | |
| 202R | HW350X350X12X19 | 490 | 0.183 | 90.078 | 81.732 | 1.102 | 11 |
| | | | | | | | |
| 207C | HW350X350X12X19 | 490 | 0.183 | 90.078 | 81.732 | 1.102 | 11 |
| | HW300X300X10X15 | 490 | 0.126 | 61.835 | 70.079 | 0.882 | 9 |
| | | | | | | | |
| 208C | HW428X407X20X35 | 490 | 0.384 | 188.41 | 96.220 | 1.958 | 16 |
| | HW300X300X10X15 | 490 | 0.126 | 61.835 | 70.078 | 0.883 | 9 |

**Table 2: Fire Endurance Times of Unprotected Columns**

| Area | Beam Dimensions | Density of steel (lb/ft3) | Area of Shape (ft2) | W (lb/ft) | D (in.) | W/D | Fire Endurance Time (min) |
|---|---|---|---|---|---|---|---|
| 202J | HM390X300X10X16 | 490 | 0.142 | 69.515 | 65.354 | 1.064 | 11 |
| | HN400X200X8X13 | 490 | 0.088 | 43.207 | 54.488 | 0.793 | 9 |
| | | | | | | | |
| 202A | HW200X200X8X12 | 490 | 0.067 | 32.743 | 38.740 | 0.845 | 9 |
| | | | | | | | |
| 202C | HN350X175X7X11 | 490 | 0.066 | 32.416 | 47.677 | 0.679 | 8 |
| | | | | | | | |
| 202F | HM390X300X10X16 | 490 | 0.142 | 69.515 | 65.354 | 1.064 | 11 |
| | HN400X200X8X13 | 490 | 0.088 | 43.207 | 54.488 | 0.793 | 9 |
| | | | | | | | |
| 202G | HM340X250X9X14 | 490 | 0.106 | 51.7305 | 55.591 | 0.931 | 10 |
| | HM294X200X8X12 | 490 | 0.075 | 36.709 | 46.142 | 0.795 | 9 |
| | | | | | | | |
| 202H | HN350X175X7X11 | 490 | 0.066 | 32.416 | 47.677 | 0.679 | 8 |
| | | | | | | | |
| 202K | HN350X175X7X11 | 490 | 0.066 | 32.415 | 47.677 | 0.679 | 8 |
| | | | | | | | |
| 202L | HN450X200X9X14 | 490 | 0.101 | 49.568 | 58.346 | 0.849 | 9 |
| | HM390X300X10X16 | 490 | 0.142 | 69.515 | 65.354 | 1.0636 | 11 |
| | | | | | | | |
| 202R | HM390X300X10X16 | 490 | 0.142 | 69.515 | 65.354 | 1.064 | 11 |
| | | | | | | | |
| 207C | HM294X200X8X12 | 490 | 0.075 | 36.709 | 46.142 | 0.796 | 9 |
| | | | | | | | |
| 208C | HM390X300X10X16 | 490 | 0.142 | 69.515 | 65.354 | 1.064 | 11 |
| | HM294X200X8X12 | 490 | 0.075 | 36.709 | 46.142 | 0.796 | 9 |

**Table 3: Fire Endurance Times of Unprotected Beams**

The above tables show that the minimum fire endurance time for both unprotected columns and beams is 8 minutes. Therefore, these structural members can withstand a fire up to a minimum of 8 minutes without considering any other control measures such as firewater cooling. Further, some structural steel members have an endurance time ranging from 10 to 16 minutes.

# 6 Fireproofing Estimation

The above fire endurance/resistance times have been calculated without taking credit for fire water cooling.

Based on the calculated fire endurance times, the pool fire scenarios have been selected for fireproofing with the criteria below:

- Pool fires with a duration of 10min and above; and
- Pool diameter of 4 m and above.

With the above criteria, there are a very few scenarios left with a pool diameter of less than 4m and these scenarios have a pool diameter of less than 0.6m. Therefore, the pool diameter criteria did not make a significant impact on selection of the fireproofing scenarios.

The fireproofing extension for the selected scenarios have been considered as described below

- Horizontal extension will be the pool diameter. However, the horizontal extension will be adjusted based on ground conditions such as trenches, dikes, sloping etc.
- Vertical extension will be the flame length.

A snapshot of the qualified scenarios for fireproofing are tabulated below:

| Isolatable Section | Release Rate [kg/s] | Leak Duration [s] | Total Inventory Released [kg] | Weather | Pool Diameter [m] | Flame Length [m] | Pool Fire Duration [s] | Qualified for Fire Proofing | Within a FPZ | Fireproofing Required | Remark |
|---|---|---|---|---|---|---|---|---|---|---|---|
| 2V-1413 | 3.8 | 57 | 218 | D 5m/s | 0.4 | 1.1 | 65142.3 | N | | | |
| 2E-1504 | 32.0 | 3397 | 108622 | D 5m/s | 7.2 | 14.9 | 38074.4 | Y | | | |
| 2E-1503 | 51.8 | 347 | 17988 | D 5m/s | 4.3 | 10.5 | 17281.4 | Y | | | |
| 2V-1412 | 1.9 | 58 | 111 | D 5m/s | 0.6 | 1.6 | 12685.7 | N | | | |
| 2V-1602 | 42.2 | 361 | 15216 | D 5m/s | 5.0 | 11.5 | 11038.9 | Y | | | |
| 2V-1404 | 45.5 | 1199 | 54576 | D 5m/s | 8.9 | 23.7 | 7442.4 | Y | | | |
| 2V-3401 | 20.8 | 2997 | 62303 | D 5m/s | 23.1 | 21.8 | 4246.5 | Y | | | |
| 2V-2002A_LPF | 37.0 | 3600 | 133363 | D 5m/s | 18.9 | 40.3 | 3969.5 | Y | | | |
| 2V-2001A_LPF | 22.0 | 2624 | 57628 | D 5m/s | 18.9 | 29.1 | 2920.6 | Y | | | |
| 2P-1703 | 34.2 | 101 | 3467 | D 5m/s | 4.2 | 14.1 | 2095.3 | Y | | | |
| 2E-1404A | 57.1 | 1496 | 85359 | D 5m/s | 20.9 | 43.2 | 2077.7 | Y | | | |
| 2E-2001 | 1.4 | 20 | 28 | D 5m/s | 0.6 | 2.5 | 1647.1 | N | | | |
| 2V-1706 | 22.2 | 322 | 7140 | D 5m/s | 7.2 | 20.5 | 1471.3 | Y | | | |
| 2V-1604 | 10.4 | 1720 | 17967 | D 5m/s | 11.8 | 28.1 | 1437.0 | Y | | | |
| 2E-1403 | 55.7 | 1036 | 57666 | D 5m/s | 21.7 | 44.4 | 1295.6 | Y | | | |
| 2V-1603 | 23.8 | 540 | 12865 | D 5m/s | 11.3 | 26.9 | 1159.0 | Y | | | |
| 2V-2001A | 22.0 | 2624 | 57628 | D 5m/s | 32.8 | 42.7 | 967.8 | Y | | | |
| 2E-1402 | 54.6 | 687 | 37555 | D 5m/s | 22.0 | 44.7 | 826.4 | Y | | | |
| 2V-1502 | 51.2 | 510 | 26142 | D 5m/s | 23.9 | 34.3 | 823.9 | Y | | | |
| 2PK-1602 | 39.9 | 86 | 3434 | D 5m/s | 8.2 | 19.5 | 689.9 | Y | | | |
| 2V-2401 | 2.4 | 872 | 2116 | D 5m/s | 6.8 | 16.1 | 687.2 | Y | | | |
| 2E-1608 | 10.6 | 199 | 2116 | D 5m/s | 6.9 | 16.3 | 663.6 | Y | | | |
| 2V-2002A | 37.0 | 3600 | 133363 | D 5m/s | 46.2 | 75.0 | 663.1 | Y | | | |
| 2E-1704 | 37.1 | 89 | 3309 | D 5m/s | 7.7 | 21.3 | 605.4 | Y | | | |
| 2P-1602A | 14.0 | 149 | 2093 | D 5m/s | 7.3 | 17.4 | 558.6 | N | | | |
| 2P-1501 | 22.0 | 121 | 2650 | D 5m/s | 9.6 | 18.1 | 525.3 | N | | | |
| 2E-1506 | 51.2 | 43 | 2199 | D 5m/s | 9.4 | 17.9 | 451.7 | N | | | |
| 2E-1708 | 36.8 | 82 | 3029 | D 5m/s | 18.9 | 19.0 | 308.7 | N | | | |

**Table 4: A Snapshot of the Qualified Scenarios for Fireproofing**

# 7 Conclusions

The pool fire modeling has been performed for a 50mm release for all the equipment handling flammable liquids. The pool fire characteristics such as pool diameter, pool fire duration and flame lengths have been estimated. Fire endurance/resistance times have also been calculated for unprotected structural steel members that are present in the areas where the pool fire potential equipment is existing.

LNG and liquefied gases used as refrigerants (propane and ethylene) are creating pool fires, however, these pool fires are not lasting for more than 10 minutes, and some fires are not even lasting for more than a few minutes. Moreover, it has been estimated that the structural steel members have a fire endurance/resistance time of a minimum 8 minutes. Therefore, the structural supports which are exposed to these short duration fires do not require fire proofing.

As a result, the pool fire scenarios which have a pool fire duration of less than 10 minutes have been excluded from the fireproofing taking credit for the fire endurance time, firewater cooling and drainage system. Overall, the risk-based approach has reduced fireproofing application from a surface are of 25406.20 $m^2$ to 17061.76 $m^2$ which is equivalent to a 32% reduction of fire proofing material.

This study has shown how calculated pool fire size and duration will play a significant role in determining fireproofing requirements, and how a risk based fireproofing design together with the calculated time to failure of structural steel members may result in a significant reduction of fireproofing material quantities, and consequently a cost saving, which is desirable in the current economy.

## 8 References


1. Arumugakan Thuraiswamy, J. B., Girada, R., Nicotra, A., "Fire Proofing Cost Reduction Measures for LNG Plant", 2017 Spring Meeting and 13th Global Congress on Process Safety, American Institute of Chemical Engineers, San Antonio, Texas, 2017.
2. J.D. Marx and A. Nicotra, "Is a Two-Inch Hole Adequate for Siting Study?", 12th Global Congress on Process Safety, American Institute of Chemical Engineers, Houston, Texas, 2016.
3. Society of Fire Protection Engineers (SFPE), "Handbook for Fire Protection Engineering", Volume 2, 5th Edition, 2016.
4. API 2218. "Fireproofing Practices in Petroleum and Petrochemical Processing Plants", American Petroleum Institute, 2013.

# 640070 Experimental and Theoretical Study on the Bund Overtopping Process after a Catastrophic Failure of LNG Storage Tanks

**Abstract:**

LNG is usually stored in large tanks, and the tank volume is up to 230,000 cubic meters. Regulations mandate the Liquefied Natural Gas (LNG) storage tanks to be equipped a bund to control the spilled liquid after a tank failure. The capacity of the bund is usually designed to be 110% of the tank volume, but the 110% rule has been proven to be ineffective to contain all the spilled liquid in the case of a catastrophic tank failure. Therefore, the spilled LNG will overtop the bund, resulting in severe consequences. Previous work mainly used water to study the effect of radius and height of circular bunds on overtopping fraction, other key design parameters of the bund system were barely studied, such as other bund shapes rather than circle, inclination angle of bund walls, and break water. The numerical studies developed Computational Fluid Dynamics (CFD) models of bund overtopping, but the boiling phenomena of LNG was ignored. In this work, the overtopping study was further enhanced by investigating the shape, the inclination angle, and the breakwater of the bund using an in-house apparatus. The numerical study integrated a boiling model into the validated CFD model, which can simulate the LNG boiling phenomena during the bund overtopping process. The findings of this work will help to prevent the bund overtopping of a catastrophic tank failure accident, and provide scientific recommendations on the bund design of the LNG storage tanks.

*Corresponding author: Bin Zhang, bzhang@njtech.edu.cn

# 640210 Multi-Objective Transportation Route Optimization for Hazardous Materials Based on GIS

**Abstract:**

Road transportation of LNG is an important public safety issue. Improper operations on LNG transportation could pose serious health risks to humans and many other living organisms, huge economic losses to industries, as well as dangerous pollutions to local environment. In addition, many unexpected and uncontrollable factors could escalate disasters induced by LNG accidents. How to effectively plan and reduce the impact caused by hazmat transportation accidents has become a major issue that scholars pay close attention to today. Therefore, it is necessary and essential for us to study on hazardous chemicals transportation route optimization. It is worth noting that comprehensive factors should be considered when routing for hazmat transportation in order to satisfy all parties' concerns. In our study, we presented a constrained multi-objective optimization model to solve this problem. Compared to the bi-level programming model, the constrained multi-objective optimization model is able to separate risk and cost, and consider additional impact factors. We used the map provided in Geographical information systems (GIS) as a basis, and use its data, such as the population density, the distance of the motorways, the location of buildings, to prove that our model is worked. We use a motorway network in Jiangsu province as a case study, which has high reliability and reference value to show the results of the framework developed.

# 639358 Transitioning Energy Systems and Shell Technologies


**Abstract:**

We need energy for just about everything we make and do. Currently energy is primarily produced using fossil fuels, however in recent years, share of renewables in energy mix has been gaining momentum, particularly in power sector. Renewables will play a key role in achieving net-zero emission as set in the Paris agreement. Note that the transitions in energy sector have happened before, where we saw growth in the use of coal, and then oil and natural gas and more recently in new technologies such as solar and wind power. This talk will review the energy systems, its evolution, and the technology development within Shell preparing for the transition – particularly in gas conversion and its utilization – example Green and Blue H2, CCUS, Dense Energy Carrier and alternate ways of synthesizing syngas.

# 639030 Conceptualization, Experimental Proof, and Life Cycle Assessment of Cargen™, a Novel $CO_2$ Utilization Technology

**Abstract:**

Dry Reforming of methane (DRM) has attracted significant scientific attention due to its potential for effective greenhouse gas (methane and carbon dioxide) conversion to 'syngas.' Syngas is a mixture of carbon monoxide and hydrogen and is used to produce a variety of chemicals, additives, and synthetic fuels. The commercial implementation of DRM is, however, limited due to three primary process challenges, *(a) Carbon formation leading to catalyst deactivation, (b) Syngas ratio adjustment requirement for downstream application, (c) high energy requirement needed to drive the reaction.* Our research group has developed and pioneered a novel technological solution, CARGEN™ technology, addressing DRM challenges. The CARGEN™ technology presents a unique and breakthrough advantage of co-producing Carbon Nanotubes (CNTs) as a co-product along with Syngas at an adjustable ratio that can meet downstream process requirements.

Some of the key highlights of CARGEN™ technology's demonstrated performance are: *(a) Ability to convert at least 65% $CO_2$ at 50% less energy compared to standalone DRM process. (b) Co-production of CNTs and Syngas, which are products of high value and demand. (c) At least 40% reduction in $CO_2$ footprint and operational costs compared to commercial reforming technologies. (d) A possibility to integrate into existing gas processing infrastructure with an attractive economic and sustainability incentive.*

In this presentation, we present the pioneering journey of CARGEN™ technology from its theoretical conceptualization to experimental proof, and to the scale-up from micro-grams of CNT production to multi-grams. The talk will also highlight the results of life cycle assessment (LCA) and a preliminary techno-economic evaluation to demonstrate the sustainability and economics benefits that can be achieved upon implementation of CARGEN™ technology.

This work was a part of a mega-project (NPRP-X 100-2-024), and Responsive Research Seed Grant (RRSG) focused on addressing DRM challenges from micro to macro scale.

References:

1. Challiwala MS, Choudhury HA, Wang D, El-Halwagi MM, Weitz E, Elbashir NO. A novel CO2 utilization technology for the synergistic co-production of multi-walled carbon nanotubes and Syngas. *Sci Rep*. 2021;11(1):1417. doi:10.1038/s41598-021-80986-2
2. Challiwala MS, Choudhury HA, Sengupta D, El-Halwagi MM, Elbashir NO. Turning CO2 into Carbon Nanotubes. *Chem Eng Prog*. October 2020:9-10. aiche.org/cep.
3. Elbashir NO, Challiwala MA, Sengupta D, Mahmoud EM. System and method for carbon and syngas production. April 2020.
4. Challiwala M, Afzal S, Choudhury HA, Sengupta D, El-halwagi M, Elbashir NO. Alternative pathways for CO2 utilization for enhanced methane dry reforming technology. In: *Advances in Carbon Management Technologies*. ; 2019.
5. Challiwala M, Choudhury HA, Elbashir NO. Regeneration & Activation of Dry Reforming of Methane Catalysts Using CO2. 2019.

# 639034 The Impact of Catalyst Synthesis Techniques on the Activity and Stability of Ni/Cu Bimetallic Systems for the Dry Reforming of Methane (DRM) Reaction

**Abstract:**

Dry Reforming of Methane (DRM) process has garnered significant attention from both industry and academia due to its potential to produce ultra-clean fuels and value-added chemicals from natural gas and $CO_2$, the major greenhouse gases. Nickel based catalysts are generally utilized for commercial methane reforming processes. However, these catalysts when used for the DRM reaction suffer intense catalyst deactivation due to coking, which is a subject of several recent studies. In our previous studies[1-3], we demonstrated the use of bimetallic catalysts comprising of nickel and copper to address the DRM catalyst deactivation. Several techniques could be used to synthesize bimetallic catalysts, which will be discussed in this presentation. Specifically, two studies will be presented, *a) the effect of doping the nickel catalyst with different loadings of copper (low, medium, and high), b) various synthesis techniques e.g., incipient-wetness-impregnation (IWI), sol-gel (SG), and co-precipitation (CP) and their impact on catalyst performance.*

All the catalysts were characterized using a variety of characterization techniques (ICP, BET, $H_2$-TPR, chemisorption, XRD, XPS, SEM, TEM, EDS, TGA, TPO, DSC). These catalysts were then tested for activity and stability over both short time on streams (<10 hours) and long time on streams (up to 70 hours) operation. In corroboration from the observations in previous theoretical studies, our experiments demonstrated that medium copper loading catalyst synthesized using IWI technique provided the best activity and stability. Interestingly, the same loading resulted in adverse effect when SG and CP methods were used. We hypothesize this affect is due to the differences in the copper distribution on the catalyst surface during synthesis procedures. In essence, this work highlights the critical role of synthesis techniques on the performance of the DRM catalysts.

This work was conducted as part of the (NPRP-X 100-2-024) project funded by the Qatar National Research Fund (QNRF).

References:

1. Chatla, Anjaneyulu, et al. *Fuel*308 (2022): 122042.
2. Omran, Ahmed, et al. *Catalysts*9 (2020): 1043.
3. Chatla, Anjaneyulu, et al. *Applied Catalysis A: General*602 (2020): 117699.

# 639320 Sustainable Natural Gas Converting Clusters

**Abstract:**

Natural gas is one of the primary energy sources in the world. Its abundance aids in meeting the global industrial demand for energy. Natural gas is primarily used to produce electricity and heat but is also utilized to produce value-added products such as synthetic crude, ammonia, methanol and hydrogen. Even though it is considered a cleaner burning-hydrocarbon compared to fossil fuels, carbon emissions released from the combustion of natural gas are still significant. This poses a major challenge to the natural gas sector especially as countries diversify away from fossil fuels (IEA, 2021). The use of renewable energy sources and resource integration methods can be used for emission reduction and to encourage sustainability. This includes the application of Carbon Capture Utilization and Storage (CCUS) and converting $CO_2$ to value added products. This work implements a multi-objective integration model to optimize a multi-product cluster in terms of the economic, emission footprint, and resource conservation. The cluster investigated consisted of Fischer Tropsch (F-T) synthetic crude production, methanol, ammonia, hydrogen and urea in addition to power and heat production from both natural gas and solar energy. The economic performance was quantified by the total profit, the emission footprint by the total amount of carbon dioxide emitted from the cluster, and the resource conservation was optimized by minimizing the total water consumption. A pareto surface was generated that shows the trade-off between the three objectives.

References:

IEA, (2021). Net Zero by 2050: A Roadmap for the Global Energy Sector. International Energy Agency

# 639077 Progress Towards a Comprehensive Model for Mass Transfer and Hydraulics for Structured Packings in Carbon Capture Service


**Abstract:**

Rate-based calculations theoretically offer more rigor and reliability for assessing column performance than the traditional equilibrium-stage approach. Unfortunately the rate-based method is ultimately tied to equipment performance correlations with limited predictive capability. We have undertaken an effort to develop significantly improved correlating expressions for mass transfer in columns containing sheet metal structured packings. To accomplish this, we have collected and simultaneously analyzed HETP data from 443 distillation experiments (Bennett, et al., 1992; Bravo, et al., 1985; Fitz, et al., 1999; Gualito, et al., 1997; and others) and 410 mass transfer area experiments (Tsai, 2010). The correlating expressions for the mass transfer performance for these sheet metal structured packings are shown below:

$Sh_V = 0.333\ Re_V^{\,1}\ Sc_V^{\,1}\ [\cos(\theta)/\cos(\pi/4)]^{0.552}$

$Sh_L = 0.1666 Re_L^{\,0.8}\ Sc_L^{\,0.758}\ [\cos(\theta)/\cos(\pi/4)]^{0.747}$

$a_m/a_d = 0.8093\ Re_V^{\,-0.0293}\ We_L^{\,0.175}\ Fr_L^{\,-0.1208}$

$Sh_V = (k_y d_e/(c_V D_V))$

$Sh_L = (k_x d_e/(c_L D_L))$

$d_e = 4\varepsilon/a_d$

The table below compares HETP predictions from the correlation above and from other popular mass transfer correlations in the open literature (Bravo, et al., 1985; Hanley and Chen, 2014; Wang, et al., 2016) with some of the experimental data we have collected. We find that the correlation of this work predicts the experimental HETP dataset with an average difference of 9.4% and a maximum difference of 50%; the next best correlation - that of Wang, et al. - predicts these same data with an average difference of 20% and a maximum difference of 172%.

| | | Pressure | Packing | Fs | HETP | This Work | Hanley & Chen | Wang, et al. | BRF85 |
|---|---|---|---|---|---|---|---|---|---|
| **Light Key** | **Heavy Key** | **Pa** | | **$Pa^{1/2}$** | **m** | **m** | **m** | **m** | **m** |
| P-XYLENE | O-XYLENE | 2000.0 | Mellapak 250Y | 0.8478 | 0.3615 | 0.4016 | 0.4318 | 0.4002 | 0.2549 |
| P-XYLENE | O-XYLENE | 13000.0 | Mellapak | 0.2332 | 0.3493 | 0.3587 | 0.3306 | 0.2851 | 0.2154 |

| | | | | | | | | | |
|---|---|---|---|---|---|---|---|---|---|
| | | | 250Y | | | | | | |
| P-XYLENE | O-XYLENE | 100000.0 | Mellapak 250Y | 0.2817 | 0.3221 | 0.3195 | 0.2677 | 0.2463 | 0.2303 |
| IC4 | NC4 | 690000.0 | Mellapak 250Y | 1.1402 | 0.2868 | 0.3061 | 0.2346 | 0.3007 | 0.3002 |
| IC4 | NC4 | 1140000.0 | Mellapak 250Y | 0.7191 | 0.2778 | 0.2773 | 0.2112 | 0.2525 | 0.2131 |
| CYCLOC6 | NC7 | 33000.0 | Mellapak 250Y | 3.1962 | 0.4111 | 0.4080 | 0.4068 | 0.4906 | 0.5029 |
| CHLORBZ | ETHBZ | 96000.0 | Mellapak 250Y | 1.5976 | 0.3548 | 0.3387 | 0.2698 | 0.3305 | 0.3636 |
| CHLORBZ | ETHBZ | 96000.0 | Mellapak 350Y | 2.0988 | 0.2631 | 0.2571 | 0.2204 | 0.2335 | 0.2671 |
| CHLORBZ | ETHBZ | 5000.0 | Mellapak 500Y | 1.0143 | 0.2195 | 0.2141 | 0.2443 | 0.1694 | 0.1196 |
| 2ME2BUTL | ISOBUTL | 20000.0 | Mellapak 500Y | 1.2422 | 0.2805 | 0.2364 | 0.4578 | 0.2673 | 0.1584 |
| 2ME2BUTL | ISOBUTL | 10000.0 | Mellapak 500Y | 1.1324 | 0.2694 | 0.2502 | 0.6447 | 0.3234 | 0.1488 |
| 2ME2BUTL | ISOBUTL | 2000.0 | Mellapak 500Y | 1.6408 | 0.3365 | 0.2934 | 1.6509 | 0.5715 | 0.1538 |
| CHLORBZ | ETHBZ | 2000.0 | Mellapak 500X | 1.3943 | 0.2846 | 0.2811 | 0.7395 | 0.2240 | 0.1389 |
| CHLORBZ | ETHBZ | 10000.0 | Mellapak 500X | 1.3977 | 0.2780 | 0.2630 | 0.4882 | 0.1957 | 0.1578 |
| CHLORBZ | ETHBZ | 20000.0 | Mellapak 500X | 0.6351 | 0.2670 | 0.2490 | 0.4033 | 0.1654 | 0.1365 |

| | | | | | | | | | |
|---|---|---|---|---|---|---|---|---|---|
| CHLORBZ | ETHBZ | 95000.0 | Mellapak 500X | 1.9088 | 0.2434 | 0.2353 | 0.2651 | 0.1741 | 0.1931 |
| 2ME2BUTL | ISOBUTL | 10000.0 | Mellapak 500X | 1.7589 | 0.3269 | 0.3225 | 2.2484 | 0.3967 | 0.1927 |
| 2ME2BUTL | ISOBUTL | 20000.0 | Mellapak 500X | 1.3913 | 0.3035 | 0.2999 | 1.4575 | 0.3175 | 0.1865 |
| 2ME2BUTL | ISOBUTL | 95000.0 | Mellapak 500X | 1.4372 | 0.2582 | 0.2609 | 0.5831 | 0.2208 | 0.1991 |
| CHLORBZ | ETHBZ | 10265.8 | Mellapak 250Y | 1.4602 | 0.3889 | 0.3810 | 0.3434 | 0.3772 | 0.3189 |
| CHLORBZ | ETHBZ | 10265.8 | Mellapak 250Y | 2.4386 | 0.3975 | 0.3883 | 0.3448 | 0.4072 | 0.3638 |
| CHLORBZ | ETHBZ | 10265.8 | Mellapak 252Y | 3.0246 | 0.4071 | 0.3918 | 0.3454 | 0.4208 | 0.3852 |
| P-XYLENE | O-XYLENE | 13332.2 | Mellapak 252Y | 2.9485 | 0.3553 | 0.3861 | 0.3363 | 0.4136 | 0.4035 |
| CYCLOC6 | NC7 | 165474.2 | Mellapak 252Y | 1.4631 | 0.3387 | 0.3479 | 0.3027 | 0.3624 | 0.3981 |
| CYCLOC6 | NC7 | 101325.0 | Mellapak 752Y | 1.7872 | 0.2271 | 0.2028 | 0.2164 | 0.1769 | 0.1943 |
| ISOC8 | TOLUENE | 13332.2 | ISP IT | 1.4225 | 0.3105 | 0.3282 | 0.3567 | 0.3517 | |
| ISOC8 | TOLUENE | 13332.2 | ISP 4T | 1.3958 | 0.5555 | 0.6257 | 0.5648 | 0.8315 | |
| ARGON | OXYGEN | 206842.7 | Flexipac 500Y | 0.9095 | 0.1811 | 0.1784 | 0.1586 | 0.1523 | 0.1533 |
| P-XYLENE | O-XYLENE | 5332.9 | Flexipac 1.4X | 2.5618 | 0.3454 | 0.3787 | 0.7143 | 0.3428 | 0.2802 |
| | | | Flexipac | | | | | | |

| | | | | | | | | | | |
|---|---|---|---|---|---|---|---|---|---|---|
| P-XYLENE | O-XYLENE | 26664.5 | 1.4X | 2.2568 | 0.3493 | 0.3490 | 0.4606 | 0.3011 | 0.3037 |
| P-XYLENE | O-XYLENE | 159986.8 | Flexipac 1.4X | 1.5127 | 0.3150 | 0.3068 | 0.2799 | 0.2503 | 0.2824 |
| P-XYLENE | O-XYLENE | 439963.8 | Flexipac 1.4X | 1.4517 | 0.2731 | 0.2801 | 0.2105 | 0.2294 | 0.2549 |
| P-XYLENE | O-XYLENE | 5332.9 | Flexipac 1.6X | 1.2931 | 0.3531 | 0.4332 | 0.7964 | 0.3858 | |
| P-XYLENE | O-XYLENE | 26664.5 | Flexipac 1.6X | 1.6225 | 0.3835 | 0.4043 | 0.5150 | 0.3551 | |
| P-XYLENE | O-XYLENE | 159986.8 | Flexipac 1.6X | 1.1101 | 0.3581 | 0.3559 | 0.3131 | 0.2956 | |
| P-XYLENE | O-XYLENE | 439963.8 | Flexipac 1.6X | 1.0491 | 0.3302 | 0.3247 | 0.2354 | 0.2699 | |

In addition, we have used this new correlation to model the performance of several carbon capture pilot plant experiments reported in the literature (Gabrielson 2007 ; Notz et al., 2012) . Those results, along with results for the Bravo, Rocha Fair correlation of 1985 (Bravo, et al., 1985), the 2014 correlation of Hanley and Chen, and the 2016 correlation of Wang, et al., are summarized below. Again, the correlation of this work outperforms the other correlations examined.

**Gabrielson Data**

**Run 1**

| | Experimental | This work | Wang | Hanley/Chen | BRF 85 |
|---|---|---|---|---|---|
| Gas CO2 conc. Bottom (%vol) | 2.62 | 2.62 | 2.62 | 2.62 | 2.62 |
| Gas CO2 conc. Top (%vol) | 1.69 | 1.645 | 1.602 | 1.089 | 1.278 |
| Liquid CO2 loading top | 0.072 | 0.07201 | 0.7201 | 0.07201 | 0.07201 |
| Liquid CO2 loading bottom | 0.178 | 0.183 | 0.1878 | 0.2451 | 0.224 |

**Run 3**

| | Experimental | This work | Wang | Hanley/Chen | BRF 85 |
|---|---|---|---|---|---|
| Gas CO2 conc. Bottom (%vol) | 4.17 | 4.17 | 4.17 | 4.17 | 4.17 |
| Gas CO2 conc. Top (%vol) | 2.36 | 2.173 | 2.095 | 1.514 | 1.64 |

| | | | | | |
|---|---|---|---|---|---|
| Liquid CO2 loading top | 0.084 | 0.084 | 0.084 | 0.084 | 0.084 |
| Liquid CO2 loading bottom | 0.169 | 0.1789 | 0.1825 | 0.2093 | 0.2035 |

**Run 6**

| | Experimental | This work | Wang | Hanley/Chen | BRF 85 |
|---|---|---|---|---|---|
| Gas CO2 conc. Bottom (%vol) | 4.58 | 4.58 | 4.58 | 4.58 | 4.58 |
| Gas CO2 conc. Top (%vol) | 2.9 | 2.883 | 2.785 | 2.142 | 2.337 |
| Liquid CO2 loading top | 0.147 | 0.147 | 0.147 | 0.147 | 0.147 |
| Liquid CO2 loading bottom | 0.226 | 0.2272 | 0.2318 | 0.2611 | 0.2523 |

**Run 11**

| | Experimental | This work | Wang | Hanley/Chen | BRF 85 |
|---|---|---|---|---|---|
| Gas CO2 conc. Bottom (%vol) | 10.27 | 10.27 | 10.27 | 10.27 | 10.27 |
| Gas CO2 conc. Top (%vol) | 8.01 | 8.048 | 7.797 | 6.474 | 7.108 |
| Liquid CO2 loading top | 0.284 | 0.284 | 0.284 | 0.284 | 0.284 |
| Liquid CO2 loading bottom | 0.4 | 0.3945 | 0.4057 | 0.4663 | 0.438 |

**Notz, et al., Experiment 2 Data**

| | Experimental | BRF85 | Hanley/Chen | Wang | This work |
|---|---|---|---|---|---|
| GASOUT mflow (kg/hr) | 65.6 | 66.3391 | 66.3941 | 66.6300 | 66.6127 |
| GASOUT mfrac CO2 | 0.088 | 0.0750 | 0.0757 | 0.0795 | 0.0816 |
| LEANIN mflow (kg/hr) | 200 | 198.3004 | 198.3799 | 197.6807 | 197.6336 |
| LEANIN molefrac (nCO2/nMEA) | 0.308 | 0.2988 | 0.3022 | 0.2926 | 0.2899 |
| RICHOUT mflow (kg/hr) | 207.4 | 205.8422 | 205.9266 | 205.2253 | 205.1806 |
| RICHOUT molefrac (nCO2/nMEA) | 0.464 | 0.4692 | 0.4712 | 0.4553 | 0.4490 |
| CO2OUT mflow (kg/hr) | 6.14 | 7.0062 | 6.9543 | 6.6847 | 6.5384 |
| CO2OUT mfrac CO2 | 0.996 | 0.9945 | 0.9945 | 0.9945 | 0.9945 |
| WATERMU mflow (kg/hr) | 1.21 | 1.4846 | 1.5416 | 1.7765 | 1.7660 |
| RICHOUT HeatX duty (kW) | 12.4036 | 11.9129 | 11.8712 | 11.9612 | 11.9935 |
| LEANIN H2O (w/w) | 0.653 | 0.6519 | 0.6514 | 0.6524 | 0.6526 |
| LEANIN MEA (w/w) | 0.284 | 0.2864 | 0.2863 | 0.2871 | 0.2873 |
| LEANIN CO2 (w/w) | 0.063 | 0.0617 | 0.0623 | 0.0605 | 0.0600 |

Future work will attempt to incorporate improved column hydraulic and pressure drop predictions to supplement the mass transfer correlation reported in this talk.

**REFERENCES**

Agrawal R, Woodward DW, Ludwig KA, Bennett DL. Impact of low pressure drop structure packing on air distillation. Paper presented at Distillation and Absorption: IChemE Symposium Series 128. Maastricht, The Netherlands, 1992.

Bravo JL, Rocha JA, Fair JR. Mass transfer in gauze packings. Hydrocarbon Process. 1985;64:91.

Chao Wang, Di Song, Frank A. Seibert, and Gary T. Rochelle. Dimensionless Models for Predicting the Effective Area, Liquid-Film, and Gas-Film Mass-Transfer Coefficients of Packing. Industrial & Engineering Chemistry Research. 2016 55 (18), 5373-5384

Fitz CW, Kunesh JG, Shariat A. Performance of structured packing in a commercial-scale column at pressures of 0.02–27.6 bar. Ind Eng Chem Res. 1999;38:512–518.

Gabrielsen J. CO2 capture from coal fired power plants. PhD Thesis, Technical University of Denmark, 2007.

Gualito JJ, Cerino FJ, Cardenas JC, Rocha JA. Design method for distillation columns filled with metallic, ceramic, or plastic structured packings. Ind Eng Chem Res. 1997;36:1747–1757.

Hanley B, Chen C-C. Letter to the Editor. AIChE J. 2012;58:132–152.

Notz, R., Mangalapally, H.P., Hoch, S., Hasse, H., 2011. Post combustion CO2 capture by reactive absorption: Pilot plant description and results of systematic studies with MEA. International Journal of Greenhouse Gas Control 6 (2012), 84-122.

Tsai, Robert E. Mass Transfer Area of Structured Packing. Ph.D. Dissertation. 2010.

# 639431 Determining the Comparative Value of Carbon Abatement Pathways in Residential Space Heating


**Abstract:**

## Determining the Comparative Value of Carbon Abatement Pathways in Residential Space Heating

### Abstract

Residential space heating accounts for a large portion of the total annual home energy use. The main source of energy for residential space heating in the U.S. comes from natural gas but where this is not available heating oil or propane may also be used. The current use of fossil natural gas for the residential space heating sector is in need of a change to support a rapid reduction in greenhouse gas (GHG) emissions. This presentation presents the results of a project analyzing renewable energy pathways at various stages of development and commercialization with the current natural gas pathway scenario. The first of the three pathways is renewable natural gas (RNG) that is derived from biogas via anaerobic digestion of dairy manure and food waste. The second pathway involves renewable hydrogen that would be blended into the RNG. The last scenario will look into the widespread use of air source heat pumps. For all scenarios the net present value (NPV) will be calculated using techno-economic analysis methodology, which is where the carbon abatement will be represented as a revenue stream using the social cost of carbon (SCC). Both the fossil natural gas and RNG derived via anaerobic digestion pathways will have their NPVs calculated with a 2020 investment start date as they are either commercialized or currently being commercialized, whereas the renewable hydrogen and air source heat pumps NPVs will be calculated with a 2025 investment date as the projected commercialization timeline. The future year (if any) where the renewable pathway NPV surpasses the fossil pathway NPV will also be calculated. Of the three renewable energy pathways air source heat pumps potentially yield the lowest carbon abatement costs and then the renewable hydrogen blended with RNG and lastly the RNG produced via anaerobic digestion scenario.



Keywords: Carbon abatement; techno-economic analysis; investment timing; social cost of carbon; net present value; renewable natural gas; renewable hydrogen; air source heat pumps

# 639395 Cost Modeling of Desalination Technologies for the Treatment of Shale Gas Waste Water (SGWW)

**Abstract:**

The advent of hydraulic fracturing (fracking) has revolutionized the natural gas industry by facilitating the extraction of vast amounts of otherwise inaccessible gas reserves trapped in shale formations. However, this revolution introduced a host of technical and environmental challenges. The fracking process produces large amounts of waste water (SGWW) that requires management. Traditionally, this water would be disposed of by injection in deep wells. New regulations, and an increased emphasis on environmental concerns, have prompted the industry to seek alternative means of managing SGWW, including recycle or reuse of the water. This would require treating the water to improve the quality and purity, to meet the standards required for the desired application. Adding to the complexity of the challenge, the nature of SGWW varies significantly over the lifetime of a gas well, as well as its geospatial location. In order to develop optimized treatment strategies, an understanding of the water quality, the technologies used to treat SGWW, and the relationship between the two is required.

In this work we studied current commercial processes and emerging technologies that are used for shale gas wastewater treatment, and we analyzed them based on their working principles, limitations, advantages, and drawbacks. Specifically, thermal based (e.g., mechanical vapor compression, multi-stage flash distillation, multi-effect distillation) and membrane-based (electrodialysis, nanofiltration, membrane distillation, forward osmosis, reverse osmosis) processes and their applications as part of the shale gas wastewater treatment technologies have been studied. A parametric study has been carried out to identify the factors that affect the capital and operating costs of these processes. In order to establish accurate correlations to model both capital expenditure (CAPEX) and operating expenditure (OPEX) equations, data mining activities were carried out based on open literature data (including both academic and company reports). We utilized data preparation techniques to refine the most relevant published plant data, which serves as the basis for the developed empirical correlations.

In this presentation we will discuss the result of these efforts. For each SGWW treatment technology studied in this work, a multi-parametric cost estimation equation for both CAPEX and OPEX was developed. Non-linear parameter fitting strategies were utilized to reflect the nonlinear nature of the relationship between costs and the parameters used in the equation. CAPEX correlations were developed to include capacity, total dissolved solids (TDS) concentration, and water recovery for the specific technology. OPEX correlations were modeled as a function of capacity, specific energy consumption, TDS concentration, and water recovery. The compiled data and developed cost correlations provide important tools for optimization, techno-economic and performance assessment of various treatment technologies, based on the characteristics of the water to be treated. Furthermore, horizontal falling film evaporators (FTFFE) were assessed as a base-case treatment technology, and the performance metrics for the base case were estimated and compared to the other technologies studied in this work.

# 639882 High Efficiency Gas Expansion Based LNG

**Abstract:**

Typical small to mid-scale LNG production processes are often biased toward capital minimization. Given rising power costs and the availability issues associated with renewable power generation, the need exists to reconsider the techno-economic balance of LNG production. Process cycle screening and thermodynamic evaluations have resulted in the development of a novel gas turbo-expansion based process class that present a 10+% reduction in power consumption relative common commercial alternatives. The combined and simultaneous optimization of turbomachinery and process heat exchange has been shown to be critical to attaining state-of-the-art thermal efficiency. The subject processes find particular utility in peak shaving and bio-gas LNG applications.

# 639029 LNG As Renewable Energy Transport Media


**Abstract:**

In recent years, reducing CO2 emission has been the focus for addressing the concern over global warming and presumed related abnormal weather phenomena such as wildfire in California, US and Turkish, serious flooding in US, Europe and China, strong hurricanes, drought, and the melting of the permafrost of Alaska and Siberia and polar ice caps. The baseload LNG industry has already taken steps to reduce the CO2 emission by using high efficiency gas turbines instead of steam turbine systems, both for the liquefaction plant and for the power plant. Will these steps be enough to satisfy the ever-increasing social concerns, or will further reductions be required?

To respond the social concern, following ideas are presented:

- Methane, SNG, synthesis and SNG Liquefaction plant: renewable energy, i.e., wind power generation lead to hydrogen production by electrolysis. Methane is synthesized by Sabatier reaction with recycled CO2.
- Receiving Terminal and Power Generation: application of high efficiency combined cycle systems, CO2 recovery by amine process from gas turbine exhaust and CO2 is liquefied for transport.
- LCO2 Tanker: the CO2 is liquefied and recycled to SNG synthesis.

This concept would lead to higher power costs for the consumer as around 30 cents per kWh, it may look costly, however it will solve the global warming issue completely, which has been started since industrial revolution at 18$^{th}$ century in UK.

# AUTHOR INDEX

**AIChE**
120 Wall Street, FL 23
New York, NY 10005-4020

ISBN 978-1-7138-6818-7

9 781713 868187

CURRAN ASSOCIATES INC.
proceedings
.com